Tucker's TOOLBOX

Turn Construction Knowledge into Selling Tools

Jack Tucker

A Service of NAHB

Across the Country Home Builders Care

BuilderBooks™
National Association of Home Builders
1201 15th Street, NW
Washington, DC 20005-2800
(800) 223-2665
www.builderbooks.com

Tucker's Toolbox: Turn Construction Knowledge into Selling Tools
Jack Tucker

ISBN 0-86718-528-7

Cataloging-in-Publication Data available from the Library of Congress

For further information, please contact:

BuilderBooks™
National Association of Home Builders
1201 15th Street, NW
Washington, DC 20005-2800
(800) 223-2665
Check us out online at: **www.builderbooks.com**

11/01 E Design Communications™/ P.A. Hutchinson 1500

Dedication

To my wife, Kate, for making my thoughts legible; my two boys, Jack and Harry, so they don't feel left out; and our dog, Coco, who could always be found sitting at my feet during those late nights of typing.

Contents

List of Figures

About the Author

With more than 20 years of home-building experience, Jack Tucker has become an expert in all aspects of home building, successfully managing large construction projects and consistently producing homes and neighborhoods of exceptional quality. He began his home-building career in 1979 and has effectively served in various capacities from framer to construction manager and regional vice president to land developer.

Jack also is a new home construction consultant, sales trainer, speaker, and author. He has combined his knowledge of hammer and nails with guitar and song to produce "Building Wisdom: Construction Training for Sales Professionals," an innovative and informative training seminar. Drawing from real life experiences, Jack has developed a reputation for exciting and engaging seminars delivered with the aid of his acoustic guitar. He has been instrumental in reshaping sales training nationwide by giving new home sales representatives the product knowledge they need to be successful in today's competitive home building market. With his high level of energy, his vast knowledge of the home building process, and his songs, which reinforce the lessons learned, Jack's unique seminars have made him one of the nation's most sought-after new home sales trainers.

As chairman for the curriculum development of the National Association of Home Builders' Advanced Level Certified New Home Sales Professional (CSP) program, Jack developed a seminar, "House Construction as a Selling Tool," to meet the needs of its members. He has delivered presentations at major conventions nationwide, including the NAHB's International Builders' Show Super Bowl Sales Rally, the Southeast Building Conference's Sales Rally and the Atlantic Builders' Convention Sales & Marketing Brunch. Jack is a weekly guest on Baltimore's WCBM talk radio show, Your Home, Your View, answering call-in questions concerning home building, and has been featured in various publications and on television.

Acknowledgments

Special thanks to Vito Simplicio and Vanessa Kangas, each of whom ensured that I answered and addressed the basic needs and concerns of sales agents; to Ken Green, who generously gave his time and expertise in areas of sales, marketing and land development and who made sure I stayed on track; to Ed Lonergan, a home builder, who proved to be a gold mine of information; and finally, to my wife, Kate, for her invaluable comments and suggestions.

Tucker's Toolbox: Turn Construction Knowledge into Selling Tools was produced under the general direction of Jerry Howard, NAHB Executive Vice President and CEO, in association with NAHB staff members Michael Shibley, Executive Vice President, Builder, Associate & Affiliate Services; Greg French, Staff Vice President, Publications and Non-dues Revenues; Eric Johnson, Publisher, BuilderBooks; Theresa Minch, Acquisitions Editor; David Rhodes, Art and Production Director; and Toral Patel, Assistant Editor.

Introduction

The idea of sharing my construction knowledge with sales professionals started when I began hosting seminars for sales professionals across the country. Having worked in the construction industry for more than 25 years—many as a supervisor—I have witnessed first-hand the frustrations of sales professionals as a result of their inability to answer customers' questions without having to first consult the project supervisor.

What is everyday construction knowledge for the supervisor is often a gold mine of information for the sales professional. Capitalizing on the supervisor's knowledge, the sales professional is able to give accurate answers and realistic expectations to the customer, communicate more effectively with the supervisor, and better manage the customer during the construction process. Understanding the nuts and bolts of a home also will produce a more confident and capable sales professional, and ultimately, a more successful sales pro than his or her peers.

The purpose of *Tucker's Toolbox* is to assist new-home sales professionals with sales presentations by addressing customers' construction-related questions and concerns both before and during the construction of the home. This book will arm sales professionals with a basic understanding of the products they are selling: the new home and the neighborhood. In any industry, product knowledge is essential; for any sales professional, it is tantamount to success.

As a guide to the general product knowledge portion of the sales presentation, this book does not seek to explain the sales process. It offers an explanation of the various methods of construction used in building a home and does not seek to recommend one method over another. You, as the sales professional, must decide which method works best for you.

The following pages contain explanations, tools, and forms regarding home construction and are intended to enlighten and assist sales professionals rather than provide an in-depth, technical analysis of the construction process. There may be exceptions to some of the statements made in this book due to differing home building methods, codes, and practices. Keep in mind that the many different construction processes, practices, and suggestions presented in *Tucker's Toolbox* may differ slightly from state to state and from builder to builder.

1

CHAPTER

Sales Center Inventory Checklist

A good sales professional should be prepared at a moment's notice to answer both the customer's and the supervisor's questions and to enter a home under construction without disrupting the construction process. By ensuring that you have your sales resources readily available, the sales/construction process will be easier.

Based on my experience with construction-related questions posed by both sales professionals and customers, I have compiled the following list of standard sales tools that should be in every sales center.

Construction File. Make sure you have a separate construction file for each home. The folder should contain the selection sheet, change orders, special plans, and any other pertinent information you have shared with the construction department for each home.

3-Ring Binder. A 3-ring binder is ideal for storing product brochures you receive from the product vendors.

Boots. Get a few pairs of oversized rubber boots for both you and your customers to wear when walking home sites during less than favorable conditions.

Hard Hats. Keep several hard hats on hand for both you and your customers. It's a good idea to select a specific color, such as yellow, so you are easily recognizable.

Paper Booties. Paper shoe covers should be worn by anyone entering a carpeted home in order to keep the carpet clean. This method is also an indication to the customer of your respect for your product.

100-Foot Nylon, Two-Sided Tape Measure. A 100-foot tape measure is a great tool to keep on hand when walking a home site. I prefer a nylon tape measure because nylon will not crack, whereas a metal tape will crack and break. A two-sided tape is even better. It eliminates having to flip the tape over to read the measurement.

Large Golf Umbrella. Don't risk losing customers to unfavorable weather!

Tucker's Toolbox. Yes, that's right, this book. You should keep it handy and refer to it often.

Cover Sheets. The cover sheets for architecture and engineering plans list most of the requirements of the subdivision. Make sure your plans include the cover sheets. Why? Say you want to know your building setback, look on the cover sheet of the recorded plat. You may, however, discover that some rural municipalities do not require cover sheets. If so, request a list of general requirements from your municipality.

Site Plan. Use this plan for checking grades, swales, and home elevations. Generally, if your issue touches the dirt or involves grades, refer to the site plan.

Recorded Plat. Legal issues are generally indicated on the recorded plat. For example, if you want to know which homes have an easement, you would check the recorded plat.

Sediment Control Plan. This plan identifies where the sediment control devices, such as silt fences, sediment ponds, stone, etc., will be installed.

Landscape Plan. Most subdivisions require approval of a landscape plan. This plan identifies common area items, such as the entrance monument and its landscaping, as well as tree-lined street requirements.

Power Plan. This plan identifies the locations of street lights, transformers, and home electric meter locations. To order the power plan, call your local power company. Ask to be connected to the design department.

Architecture Plans. Make sure you have a complete set of the most current architecture plans, including all options.

Home Owner Association (HOA) Documents. If there are any covenants specific to your neighborhood, a set of HOA documents should have been drafted and distributed by the developer to the builder for distribution by the sales professionals to the customers.

2

CHAPTER

Teamwork

A good relationship between the sales professional and the supervisor is essential for good customer relations and for delivering a top-of-the-line product to the customer. As in any industry, there is more often than not some degree of friction between different departments, and this is not uncommon with the construction and sales departments of a home building company.

I believe that some of the friction stems from the fact that sales professionals tend to focus on the overall picture, failing to address the smaller details that go into the home, whereas the supervisors concentrate more on the specifics than on the overall picture. Also, the sales professional sees the home in a personal light whereas the supervisor looks at the home in an impersonal light; to supervisors, the home is a by-product of the schedule. Whatever the reason, there are many ways to establish an optimal working relationship between sales and construction.

The following is a brief list of rules for dealing with both the customer and the supervisor during a home's construction. Remember that the better your relationship with the supervisor, the better you will be able to handle the customer and generate more sales and referrals. Your sales presentation becomes more credible because you can set realistic customer expectations and provide accurate answers to customers' construction-related questions.

Do write a neat, accurate, and legible selection sheet.

The selection sheet is an order form that identifies the amenities and options the customer has chosen for the new home. The selection sheet, completed by the sales professional, is given to the construction department to generate an approved site plan, secure a building permit, distribute to the trade professionals, and open accounts for the home.

Because so many people read and use the selection sheet, it is critical that it be filled out neatly, accurately, and legibly. Any deviation adds building days to the home, and any mislabeling or incorrect information can be quite costly. Also, bear in mind that the selection sheet is read by the trade professionals, many of whom may have a limited knowledge of English. Therefore, avoid using abbreviations. For example, does "AL" signify "Almond," "Alabaster," or anything at all to the trade professional, who matches each letter on the selection sheet with the corresponding letters on the paint can? When it comes to kitchen appliances, washers, and dryers, always include the correct make, model number, and color.

Do create a standard feature and option list.

It is vitally important that both the sales and construction departments know the home's standard features as well as its options, and can easily differentiate between the two. When the construction department budgets a home and subsequently contracts the trade professionals and orders the materials to build the home, it does so according to the home's standard features. Any additional items are considered options, each of which must be listed. If the sales department mistakenly sells an option as a standard feature, the option will not be included in the base cost of the home and, when installed, will show as a variance against the supervisor's budget.

Do set proper customer expectations concerning the construction process.

First and foremost, "confront the confrontation." In the same way that the sales department sets the proper expectation for the loan process, it is just as critical that it set the proper expectation for the construction process. For example, have you ever been on the verge of closing a sale when another customer walks in complaining about the wrong color of shutters on his or her home? It would have been better to explain at point of sale that the shutters come from the factory primed and the selected shutter color would be applied after installation.

By setting the expectation at point of sale, everyone is happy. So make a list of issues that always seem to arise and inform the customer upfront. Items you might have on your list include:

- Shutters come from the factory primed and that the color selected for the shutters will be applied after installation.
- There is a possibility that the home may not start on time due to a delay in the permit process and/or unfavorable weather conditions.
- There is a possibility that the front door may be dented but will be fixed just before settlement in case it suffers any further denting;
- There is a possibility that a window may break but will be fixed before settlement when the glass repairer fixes all broken windows in the neighborhood at one time.
- There is a possibility of bad language or loud music on the site.
- Each trade professional tends to only take care of his or her own tasks and is unconcerned about affecting another trade.
- The pre-settlement orientation is intended to teach the customer how to operate, maintain, and care for the home and is not the time to compile a final punch list.
- The customer should be instructed on how the warranty service program works and what items are classified as homeowner maintenance.

Don't schedule work directly with the trade professionals.

The question of "who is in charge" should never be an issue for the trade professional. The trade professional's instructions should come directly from the supervisor so that the supervisor is always aware of what is being built, who is building it, and how much is being charged. Signed paperwork identifies both the trade professional's scope of work and specifies the payment he/she will receive upon completion of the work. It also authorizes payment for the work. Only the supervisor knows exactly the duties and responsibilities of each individual trade professional.

Don't allow the customer's personal trade professionals to work on the home.

By allowing an unauthorized trade professional to enter and work in a home under construction, you are only adding to the builder's list of liabilities. Does the unauthorized worker have general liability, vehicle insurance, and Workers' Compensation insurance? Who is responsible if the unauthorized trade professional is injured, injures someone else, or damages another trade professional's work? Who is responsible if an authorized trade professional damages an unauthorized trade professional's work? Who is responsible if the unauthorized trade professional does not clean up after his or her work is complete or slows down the construction schedule, ultimately postponing settlement? Finally, who is responsible if the builder cannot get a final inspection because work performed in the home requiring a permit and inspection was not permitted or inspected?

Do have a weekly construction meeting.

Building good communication is essential to the successful building of a home. When the home building process starts, the sales department submits a selection sheet to the construction department specifying the house type and options chosen by the customer. A change order for any subsequent customer change requests is submitted in the same fashion during the construction process. To ensure a smooth transition of information, it is critical that the sales and construction departments hold a structured and mandatory weekly meeting. At a minimum, the objective will be to review the selection sheets and change orders for all homes under construction and to identify any inaccuracies or items that may have changed during the preceding week. Keep in mind that these meetings are not a time for fault-finding—schedule a different meeting for that. Its sole purpose is to identify the expectations for the homes under construction.

Do have a pre-construction meeting.

For the home construction process to start smoothly, a pre-construction meeting among the supervisor, the sales professional, and the customer is essential. A pre-construction meeting ensures smooth communication of the customer's selection sheet (the home's amenities) and the change orders to the supervisor.

At the meeting, the supervisor should review the selection sheet line by line with the sales professional and customer to minimize any margin for error and to set the proper expectations. The meeting is a time for communicating the customer's selections; it is not a time for the customer to ask questions about selections. All selections should be finalized prior to the meeting. If the customer has any open issues, the pre-construction meeting should be postponed. The sales professional and customer should instead schedule a separate meeting with the construction department to address any customer issues. When all issues have been resolved, a pre-construction meeting can be scheduled.

Keep in mind that the supervisor has scheduled only enough time for a pre-construction meeting, which does not include a question and answer session.

Do have a pre-drywall meeting.

Before drywalling a home, the customer should visit the site to review all of the selections with the supervisor to ensure that everything has been completed and that the customer's expectations will be met. The pre-drywall meeting is not a time for the customer to make changes, which will undoubtedly delay the construction process.

Do communicate effectively with the supervisor.

Supervisors are more accustomed to giving orders than receiving them. After all, they are responsible for managing the construction schedule, which involves lining up the trade professionals and ordering the building materials for the home. Therefore, rather than "ordering" the supervisor to do something, you may want to "request" it.

Do solidify your relationship with the supervisor.

Anything you do to make the supervisor's life easier will earn you the supervisor's respect and cooperation. Seemingly insignificant gestures such as sweeping out a home rather than requesting that it be done for you, or presenting the supervisor with a lunch that does not come from the "roach coach" will result in a more cooperative supervisor who is also more receptive to your requests and concerns.

3

CHAPTER

Scheduling

As you walk through the home, remember that it is built one trade at a time. One trade should not begin work until the previous trade has completed 100 percent of its work and all necessary materials for that trade are on site. This method of building is referred to as "critical path scheduling."

For example, if the plumber is scheduled to install the rough plumbing (also known as "top out") and the framer has not yet built the garden tub box, then the plumber cannot finish 100 percent. Therefore, the plumber should not yet be in the home. The supervisor will spend more time trying to get the plumber back to the home to complete his work than if he had either just waited one more day before allowing the plumber to begin or made sure the framer was finished before he left. In general, the schedule should be filled out on the basis of one trade professional at a time in the home. Note that each mechanical trade professional is allotted three days to complete 100 percent of his or her work.

When I am asked to evaluate the strength of supervisors, I examine their schedules, not their job status, and walk through their homes. I also check that day's schedule to see if the scheduled trade professionals are in fact working. I evaluate the work in place to determine if all work to date is 100 percent complete, i.e., whether or not the supervisor's homes were ready for that day's scheduled trade professionals. The satisfaction of both these conditions is usually a good indication of the supervisor's strength.

To help you get a feel for the number of days each major trade takes to complete his or her work, refer to figure 3.1. Note that these figures are only estimates. If you want to use the table as a guide, use the line adjacent to the estimate to fill in the actual number of days it takes your trade professionals to complete each task. The total is the estimated number of the home's working days, not the total number of building days or calendar days, which includes weekends and holidays.

Figure 3.1 How long does each task take?

Task	No. of Days
Foundation	7-30
Frame	15
Roof	2
Paint Exterior	2
HVAC Rough	3
Plumbing Rough	3
Electric Rough	3
Second Inspections	4
Insulation	1
Siding	5
Drywall	10
First Trim	3
Exterior Concrete	3
Paint and Enamel	5
Wall Tile	2
Floor Tile	3
Vinyl Floor	1
Grills & Condenser	1
Final Plumbing	3
Final Electric	3
Rough Clean	1
Punch Out	5
Final Grade	1
Second Paint	2
Hardwood Floor	1
Sand in Place	7
Carpet	1
Final Grade	1
Landscaping	3
Paint Touch-Up	1
Final Clean	1
Sign Off	5

Tucker's Law is a formula I use to estimate the settlement date of almost any home under construction. Tucker's Law is based on three things—

- The estimated total building days
- The current stage of construction
- The current stage of construction phase number (see figure 3.1).

To estimate how many days from the current stage of construction until the settlement date, (a) identify the current stage of construction of the home, (b) identify the phase number from figure 3.2, (c) multiply the phase number by the estimated total building days, and (d) move the decimal point one place to the left. This will tell you approximately how many days until settlement.

Figure 3.2 Tucker's Law

If it typically takes a builder 120 days to build a home and the home is drywalled in 5 days:

$$120 \times 5 = 600$$

Move the decimal point one place to the left:

$$60.0$$

Then the home should settle in 60 days!

Task	No. of days
Frame	8
Shingles	7
Rough Mechanicals	6
Drywall	5
Cabinets	4
Final Mechanicals	3
Floorings	2

4

CHAPTER

Walking a Home Under Construction

It is, of course, the supervisor's responsibility to build the home and to ensure that it is built correctly and according to plan. However, from time to time, the sales professionals should "walk" the home under construction in order to answer or satisfy either their questions or the customers' questions or to reassure themselves that the home is progressing smoothly.

The difficulties and frustrations you may encounter when walking a home under construction for the first time are understandable. What comes to mind when I recall first walking homes is a feeling of intimidation that took several months to overcome. So if you are not used to walking homes—relax. After a few months, you will get the hang of it and feel a lot more comfortable.

The first thing to consider when walking a home under construction is safety. Be conscious of your shoes, baggy clothing, or loose jewelry, and always wear a hard hat. Be aware of your surroundings on the construction site.

If you notice any infractions by the trade professionals or suppliers when you are walking a home, inform your supervisor. Another set of eyes helping the supervisor to oversee the 50 trades who install more than 20,000 parts in the home is always appreciated. Nagging or criticizing, however, is not. Please approach the supervisor respectfully.

Some phases of construction are relatively easy to walk, while other phases are a little confusing. There is certain information you should know when walking your homes under construction:

Leaving the Office

Double check that the selection sheet (the construction order form) and approved change orders (addenda to the selection sheet) include all of the options and amenities your customer has chosen. Remember, these forms are passed onto the trade professionals. If something is missing from the forms, it will be missing in the field.

Check your site plan to make sure that all of the selected options that will affect the home's footprint are reflected. Depending on how your company builds, possible options to note include areaways, walkout basements, fireplaces, bay windows, box windows, stoops, long porches, garage sizes, decks, patios, service doors, basement windows, cover percentage calculations, and slab step-downs.

Stake Out

Make sure the home has been staked out per the approved site plan. Check that the home has not been flipped, moved back, or pushed forward. From the street, look at the lay of the land, the way the land falls, and whether there are any structures (fire hydrant, street light, transformer, catch basin, trees, steep slopes, etc.) that will impede the placement of the home or the driveway. Check the elevation the engineer has set for the first floor. To do this, try to imagine the actual height of the first floor.

Before I became comfortable siting homes, I used 5-foot planting stakes that I purchased from my local nursery and actually stuck them in the ground at the four corners of the home I was about to build. I then placed a piece of engineer's ribbon on the stake at the actual elevation of the first floor. It will take a few attempts to feel comfortable with what you are doing, but once you get the hang of it, it will help you visualize the elevation and location of the home.

Excavation

There isn't much to look at during this phase beyond checking that the excavated hole matches the footprint of the home and includes all approved options. Remember, though, that the excavator has purposely excavated an additional 4 feet wider on all sides of the home (referred to as the overdig) to allow the trade professionals room to work. As an example, the excavated hole for a 50-foot by 30-foot home should measure approximately 58 feet by 38 feet.

Basement Foundation

Check that all selected options are reflected in the foundation. You should visit the site three times during this stage. The first visit is at the footing stage. Make sure that the footprint of the footings matches the options selected. For example, is there a footing for the optional fireplace or two-car garage? At the second visit, check that the wall height is correct and check that the windows and/or service doors are in the correct locations. After the plumber has installed the underground plumbing rough and before the concrete is placed, visit the site a third time to double check that the plumbing for any basement options, such as a half bath, full bath, laundry sink, or wet bar, has been installed. Again, you are merely checking for options.

On-Grade Slab Foundation

You should visit the site twice during this stage. The first visit should immediately follow the installation of the form boards. Check that the options, which affect the footprint, are represented in the formwork. In some instances, not all of the form boards are installed at this stage, so you will need to get a feel for how your supervisor handles this process. The second visit should be immediately prior to placing the concrete. Check for other items, for example the plumbing for a wet bar, an island's electric wire rough-in, wires for floor plugs, or the HVAC trade professional's downdraft ductwork has been installed.

Lumber Package

After the lumber package has been delivered, there are a few things you can check. Once again, check for options. Make sure the studs in the package or in the panelized wall are the chosen length, usually 8 or 9 feet. Look at the roof trusses and check that the correct roof type was delivered. In a gable home, most of the roof trusses are identical (known as "common" trusses). A hip roof, however, will have many different trusses (used in building the hips), which gradually diminish in size at approximately 2-foot intervals. If vaulted or cathedral ceilings were selected for any of the rooms, make sure the scissors trusses are there. If the home features a library, morning room, den, or some other type of option that necessitates "bumping out" a section of the home, check that there is a group of approximately 10 trusses which are identical and smaller than most of the other trusses. Most likely, these trusses will be for that extension.

Frame

When I began building homes, I found that walking a frame prior to drywall was the most challenging stage of construction when trying to visualize the finished home. The drywall defines the rooms. Without drywall, rooms appear to blend into one another. After the framing stage, the home is essentially built, and it is then just a matter of finishing it. Remember, a home is built in two basic stages—the building stage and the finishing stage. If the finishing parts do not fit the frame, then the framing has to be re-done. For this reason, drywall is a critical stage for a supervisor. If something is amiss, the supervisor has to tear it out and re-build it. Therefore, after the framing is completed and all rough openings have been installed, take your selection sheet and walk the home, checking for elevation and window sizes, that every option has been installed and that the colors for items such as shingles, siding, and brick are correct. For an engineered fireplace, check the size and model number.

Finishing

The drywall stage takes approximately 7 to 10 days to complete. The home takes on a new appearance after the drywall has been hung. The drywall hangers end the "building" stage of the home, and the finishers can now begin the "finishing" stage of the home. It is a good idea to check that drywall has been installed on all windows and door openings that require drywall. This is also a good time to make sure that the finishers understand what finishes are required for each room's walls and ceilings. Make a note of the rooms to be wallpapered, textured, or not textured, so those areas can be properly prepared.

Because the home is now in the finishing stage, walking the home entails checking for correct colors, products, and model numbers. As you walk the home from this point until settlement, try to anticipate the installation of the following: wall and ceiling paint, wall and floor ceramic tile, cabinets, vanities, kitchen countertops, bathroom tops, vinyl flooring, fireplace mantel and finish, stair rails and finish, door knobs, bath hardware, plumbing fixtures, electric fixtures, switches and outlets, wallpaper, hardwood, carpet, flatwork layout, grading, and landscaping.

Deliveries

In addition to the delivery of the frame package, other products are delivered and usually stored in the garage. You can inspect these products prior to their installation to ensure that the correct products were delivered.

It is not necessary to count every window, but if you are walking past a group of windows that were just delivered and you notice that they are aluminum when wood was selected, inform your supervisor. As you become more familiar with window packages, you can check that the additional windows that come with certain options have been delivered. For instance, if an open foyer option was selected, make sure the large foyer window was delivered. If the front elevation calls for oval windows, make sure they are there.

When checking the brick, look for the correct color and size for both regular and oversized bricks selected. If the home will have a masonry fireplace, look for the white brick; this will let you know whether or not the optional fireplace brick was shipped.

Check that you have the correct siding, which is usually delivered and stored in the garage. Look at the trim package sitting in the garage. Again, I am not suggesting that you check every door, but if the standard is slab doors and the six-panel option was selected, you might want to verify that the doors in the garage are six-panel. When the cabinets have been delivered and stored in the garage, check the label on one of the boxes to verify that the correct cabinet style and color was delivered. Look for any noticeable differences, such as the addition of a pantry cabinet or oven cabinet, between the standard kitchen and an optional kitchen you may offer. If, for example, a long vanity top is part of an optional garden bath, was it delivered?

Again, you are merely checking for something that is non-standard, which should be easy to identify. If you cannot locate it, there is a good chance that it was overlooked. When the appliances are delivered, make sure the model numbers and colors are correct. If it is offered with the fireplaces, check for the mantels. If the fireplace front is bricked, make sure the design of the facing was correctly installed.

5

CHAPTER

Common Building Codes

All builders have to follow certain basic building codes. Familiarity with codes and explaining them to the customer will both impress the customer and set you apart from your competition. Before reviewing codes, it is important to check with your supervisor that the construction practices in your area use those codes and to subsequently scratch out any of the following codes that do not pertain to you.

Although these codes may differ slightly from state to state, listed below are the codes most commonly used nationwide.

Sitework

- Virgin soil is understood to have a soil bearing capacity of not less than 2,000 psf (pounds per square foot).
- Exterior grade starts at the home and must slope away from the home with a minimum fall of 6 inches in the first 10 feet (5 percent slope).

Concrete

- Concrete must comply with the requirements of ACI 318 (American Concrete Institute).
- In extreme frost conditions, exterior concrete must be air entrained.
- The bottom of the footing must sit below the frost line.
- Below grade slabs require interior drain tile to channel any ground water to the sump crock.
- Slabs under conditioned spaces require 6-mil poly.
- In frost conditions, the perimeter of on-grade slabs requires a 1 1/2-foot by 24-inch rigid R-8 Styrofoam beneath the concrete.
- Garage slabs must be a minimum of 4 inches below the interior fire door sill and slope towards garage door.

- Top courses and the area below any beam of a CMU (concrete masonry unit) or cinder block wall must be solid-blocked.

Masonry

- Brick ties must be installed at intervals of 16 inches vertically and 32 inches horizontally.
- Weep holes must be installed every 4 feet on center in the first course of a brick wall which abuts a conditioned space.
- The top of a chimney must be at least 10 feet from a structure and constructed 2 feet taller.

Fire & Safety

- Garage walls and the ceiling of adjoining homes without interior fire suppression must be 1-hour fire-rated.
- All sleeping areas require two means of egress. Every bedroom must have a window with a bottom sill no higher than 44 inches or less from the floor and with an operable area must be a minimum of 5 sq. ft. (first floor) and 5.7 sq. ft. (second floor), with a minimum width of 20 inches, and a minimum height of 24 inches.
- Any glass within 2 feet of a door, 60 inches from a tub or shower, or located at the bottom of a staircase must be safety-glazed.
- Natural light/ventilation requirements state that basement windows must be 2 percent/1 percent of basement floor square footage. All other windows must be 8 percent/4 percent of that room's floor square footage.
- Fire stopping or draft stopping is placed at all vertical and horizontal spaces between floors.

Ceiling Heights

- Minimum basement ceiling heights are 7 feet 8 inches below bottom joist and 6 feet by 10 inches below mid-bearing beam or any ductwork.

Lumber & Framing

- Any lumber closer than 8 inches to dirt is required to be pressure-treated.
- Any lumber that touches concrete must be pressure-treated.
- All lumber must have a grade stamp.
- All joist and header spans must not exceed lumber material specifications.
- Wall racking must be addressed with corner sheathing or corner diagonal bracing.

Load

- Minimum requirements for a load-bearing wall are a double top plate, single bottom plate, and studs spaced at a maximum of 16 inches on center.
- Minimum requirements for a non-load-bearing wall are a single top plate, a single bottom plate, and studs spaced at a maximum of 24 inches on center.

Stairs & Rails

- Interior stairways must not be less than 36 inches wide, with no less than 6 feet 8 inches of headroom, with handrails to be placed 30 to 38 inches above the nose of the tread, while not protruding more than 3 ½ inches into a 36-inch-wide stair opening.
- All stairs with more than two risers require a handrail.
- All porches with a drop greater than 30 inches require a level handrail.
- Level and rake rails must have a minimum height of 36 inches, with balusters installed so that a 4-inch sphere will not pass between them.

Ventilation & Thermal Protection

- Insulation requirements are regionally specific, but all vapor barriers must face the conditioned space.
- Kitchen exhaust fans must be a minimum of 100 cfm (cubic feet per minute); bathroom exhaust fans must be a minimum of 50 cfm.
- All roofs with an attic space of more than 30 inches high require ventilation.

6

CHAPTER

Option Task Breakdown

Sales professionals typically offer the customer a choice of several options in the home. The cost of each option is affected by the complexity of installing the option and the current construction stage. The optimal time for selecting an option is prior to the start of the home. Because customers often question the cost of the option, the following is a list of the most common options. You can use this list to provide your customer with an idea of what is involved in the installation of the option, thereby justifying its cost.

Bathroom. Framing, exhaust fan, plumbing fixtures, electric fixtures, drywall, structural drywall (green board or wonder board) for wall tile, trim, doors, cabinets and tops, tile, shelving, bath accessories, paint, flooring.

Bay Window or Box Window. Headers, extended floor system, windows, exterior trim, paint, shingles with flashing, interior trim, window sills, flooring.

Brick Front. Wider footing and foundation wall (or wider slab), brick, brick freeze board, paint, extended doorjambs.

Finished Basement. Framing, lumber, HVAC, electric fixtures, insulation, drywall, interior doors and trim, handrail, paint, flooring.

Fireplace, Engineered. Turned floor joist system, chase, header above fireplace opening, gas pipe (optional), electric and switch for blower (optional), siding, roof flashing.

Fireplace. Concrete for footings, walls and hearth, header above fireplace opening, steel lintel, fire brick, brick, chimney gas pipe (optional), electric and switch for blower (optional), roof flashing.

Garage. Concrete for footing, walls and slab, framing, lumber, roof trusses, garage door, fire-rated ("B" label) service door, exterior and interior lights, electric switches and GFCI plug, fire-rated drywall, exterior trim, roofing, paint, siding, gutters, down spouts.

Garden Bath. Enhanced floor system, tub deck and additional walls, tempered windows (safety-glazed), tub with jets, electric fixtures, larger vanities, tops, mirror, shower enclosure.

Laundry Room. Framing, dryer vent, plumbing fixtures, electric fixtures, cabinets or shelving, flooring.

Long Porch. Concrete for footings, walls and porch slab, (joist and decking material for wood decks), decking, roof, soffit, railings, columns, gutters, downspouts.

Open Foyer. Oak stairs, scaffolding for drywall, interior rails, foyer window, chandelier, paint.

Service Door. Exterior door and lock, header for wall system, exterior light fixture and switch, exterior landing, paint.

Skylight. Framing, skylight, insulation, roof flashing.

Smooth Ceilings. Labor for additional skim coats, paint.

Sunroom or Library. Concrete for footings, walls and basement floor (or slab), framing, lumber, roof trusses, windows, doors, upgraded HVAC system, insulation, drywall, window sills, electric fixtures, exterior trim, roofing, paint, siding, gutters, downspouts, trim, paint.

Vaulted Ceiling. Roof framing or trusses, drywall, batt insulation.

Walk-Out Basement. Concrete for frost footing and landing, 2-foot rigid insulation, exterior door and lock, header with joist hangers, exterior light fixture and switch, exterior trim, paint.

Walk-Up Basement. Same as walk-out basement, plus concrete for walls and steps, drain for landing, iron level rail and hand rail.

7

CHAPTER

Pre-construction Essentials

There are a variety of matters that come into play prior to the actual construction of the home. If the home is to be a part of a neighborhood, then the neighborhood project has first to be designed and initiated. A number of factors have to be considered and tasks performed regarding the development of the land upon which the home will sit. Listed below are several considerations that have to be taken into account before the construction of the home begins, along with several frequently asked questions regarding the home's "pre-construction" phase.

What is a building code?

A building code is a written document used by a municipality. It defines the regulations governing the construction and occupancy of a building. Although the International Residential Code has been adopted by many states, there currently is no national building code. Until one is adopted nationally, local municipalities dictate the codes chosen to govern the building standards unique in their area.

Each codebook is essentially the same—they differ only in their governing of conditions in their areas. For example, Florida codes will include hurricane-proof measures, whereas California codes include earthquake protection measures. IRC (International Residential Code), BOCA (Building Official and Code Administrators), CABO (Council of American Building Officials), IRC (International Residential Code), UBC (Uniform Building Code), ICBO (International Conference of Building Officials), SBCCI (Southern Building Code Congress International) and SFSBCCI (Southern Florida Southern Building Code Congress International) are the principal codes used throughout the United States. When drawing an architecture plan, it is important that the architect is aware of the building code adopted by the municipality governing that area.

What does the term "structural integrity" mean?

Structural integrity means building the home exactly as the architect has designed. It is ensuring that loads, such as gravity and wind, do not compromise the soundness of the home.

Why do construction delays occur?

The construction of a home can be delayed for many reasons, including bad weather, trade professional unavailability, supplier unavailability, supervisor inexperience, rework, building a new product, bad soils, unexpected ground water, municipal regulations, miscommunication, incomplete selection sheets, change orders, inaccurate special request descriptions, customers interfering with the trade professionals, or allocating resources to a particular month's settlements.

Is it advisable to use a scale?

In my opinion, scales should not be allowed in the sales center, even though construction scales are necessary from time to time. However, if there is a large margin of error, then there is generally no problem in using a scale. Before you use a scale, check to see that you are not using an engineer's ruler (tenths of an inch) on the architecture plans or an architecture ruler (whole inches) on the site plans.

Also, store your scale away from direct sunlight, which could cause it to expand, or an air conditioner, which could cause it to contract. Expansion and contraction of the scale could make critical dimensions inaccurate. Find out if you are using a rack set of plans or a reduced set because a reduced set of plans is not to scale. And remember that a crease in a plan could actually add an inch or two to a measurement. Finally, consider how some revisions are made to plans. Sometimes only the measurements are changed while the line stays on the plan in the original position. If you scale from that line, inaccuracies appear because that portion of the plan is no longer drawn to scale.

What is the best way to reorder plans?

Make sure you keep all your construction plans up to date. When ordering a new set of plans to replace an assumed outdated set, always order them by the latest revision date. If the revision date has changed, your present set of plans is not current. If it has not changed, your plans are current. Information on revision dates appears in the title block at the bottom of the page.

What are the ramifications of a change order?

When home construction starts, all trade professionals and suppliers receive a selection sheet that identifies the model chosen and options selected. When a change order is issued, the new information must be relayed to the trade professionals and suppliers who are affected by the change order, usually resulting in some confusion in the field. Of utmost importance is that each affected trade professional and supplier understands exactly what is expected.

A change order not only slows down the building of the home, but it also affects other homes under construction. The trade professionals are pulled from their current projects to return to an already completed home to satisfy the change order. The supervisor is actually the one who suffers because although the trade professionals are paid for their work on the change order, the supervisor is not. Furthermore, because the number of building days increases for the home with the change order and the homes from which the trade professionals were pulled, the supervisor may lose some or all of his bonuses due to the increased building days. As a result, try to discourage change orders and encourage the customers to select options at the point of sale. Remember, adding a bay window to a home that has been drywalled is actually more difficult and frustrating than building an entire home!

Who is affected by a bay window change order?

The following is a list of people/trades/vendors who must be contacted in order to update the selection sheet in the event of a bay window change order:

Sales professional
Sales Manager, Project Manager
Purchasing Manager, Production Administrator
Permit Administrator
Supervisor
Engineer
Foundation
Lumber Yard
Framer
Window Supplier
Window Installer
Roofer
Cornice
HVAC
Electrician
Insulation
Drywall
Interior Trim
Painter
Flooring
Cleaner
Surveyor

NOTE: Because of the number of people involved in a bay window change order, the likelihood of an error being made increases dramatically.

What are the ramifications of custom requests?

Prior to writing a custom request, make sure:

- The customer knows exactly what he or she wants.
- You understand exactly what the customer wants.
- You communicate to the supervisor exactly what the customer wants.

If both you and the customer are unsure, then in all likelihood, the custom request will be incorrectly installed. Just remember that the customer knows what he or she doesn't want once it is installed. For custom requests, I suggest taking two pieces of paper: one for you and one for the customer. Have the customer describe and sketch a diagram of exactly what he or she wants, and you do the same. Then compare notes.

For example, if the request is for an additional door, make a sketch of exactly where the door will be installed based on your sketch and the customer's sketch. Include the door's size, style, location, swing, paint type, and color. Also, indicate whether or not light switches are needed at the location of the door. In other words, eliminate any open issues by providing as much detail as possible. Finally, have the customer sign the completed sketch.

What is a prorated warranty?

A prorated warranty means that the homeowner can receive reimbursement for only the remainder of the life of the product. For example, if a home has a 20-year prorated shingle warranty on shingles that cost $2,000 when they were originally installed and the roof fails after 19 years, the manufacturer would reimburse the homeowner only $100 (i.e., 2,000 x [20-19]/20).

What is a dye lot?

Vinyl flooring, brick, paint, carpet, tile, or basically anything colored will have a dye lot. For example, when paint is produced, the ingredients that make up the paint are mixed together. Some of these ingredients are natural products, and the next time a batch of the same color paint is produced, the ingredients will differ slightly in color. This results in a paint batch that is not 100 percent identical to its predecessor.

This same concept applies to other colored materials in the home. For example, when vinyl flooring is installed, the supervisor often puts aside a scrap of the vinyl in case the floor needs to be patched before settlement. By ordering a piece of replacement vinyl to repair a tear, there is a good possibility that the color of the new piece of vinyl will be a different shade than the original because it was produced in a different dye lot.

How is the builder's master key system terminated?

Some builders often adopt a builder's key system for their homes so that one key will allow access to all homes under construction. When the homeowner inserts his or her house key into the lock for the first time, it breaks the builder's key system for that particular lock and prevents the builder's key from working. When homeowners assume occupancy, it is important that they insert their keys into every exterior lock in the home in order to break the builder's key system for each lock. This ensures the homeowner's security.

What is the difference between architects' scales and engineers' scales?

When architects design building plans, they measure in inches. When engineers draw site plans, they measure in tenths of an inch. Architects typically use a $^1/_4$ scale where $^1/_4$ inch denotes 1 foot. Engineers typically use a 40 scale where 1 inch denotes 40 feet.

What is control?

Control consists of two things: location and elevation. A final survey will contain notations such as N428,316.97-E890,802.01 and N428,583.41-E890,621.42, which refer to two distinct spots on the earth. The earth is mapped out in a grid consisting of lines of longitude and latitude (see Figure 7.1); N428,316.97-E890,802.01 and N428,583.41-E890,621.42 identify two locations on the earth. These are the actual locations on the earth of the two front corners of the home site. The elevation, or the height of the basement and/or first floor of the home, is based on ground zero, which is the mean of the high and low tides. In Figure 7.2, the survey's tripod has a level line elevation reading of 24, i.e., 24 feet above sea level. Because the level line has an elevation of 24 feet and the top of the slab is 4 feet lower than the level line, we can deduce that the slab is at an elevation of 21 feet. Coastal regions will typically have elevations of 100 feet or less, whereas homes in the Rocky Mountains could have elevations as high as 10,000 feet.

Figure 7.1 Longitude and latitude of earth

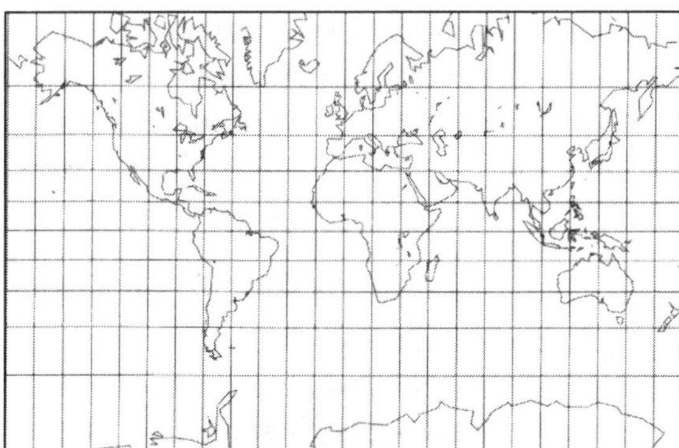

Figure 7.2 Surveyor shooting 25-4=21

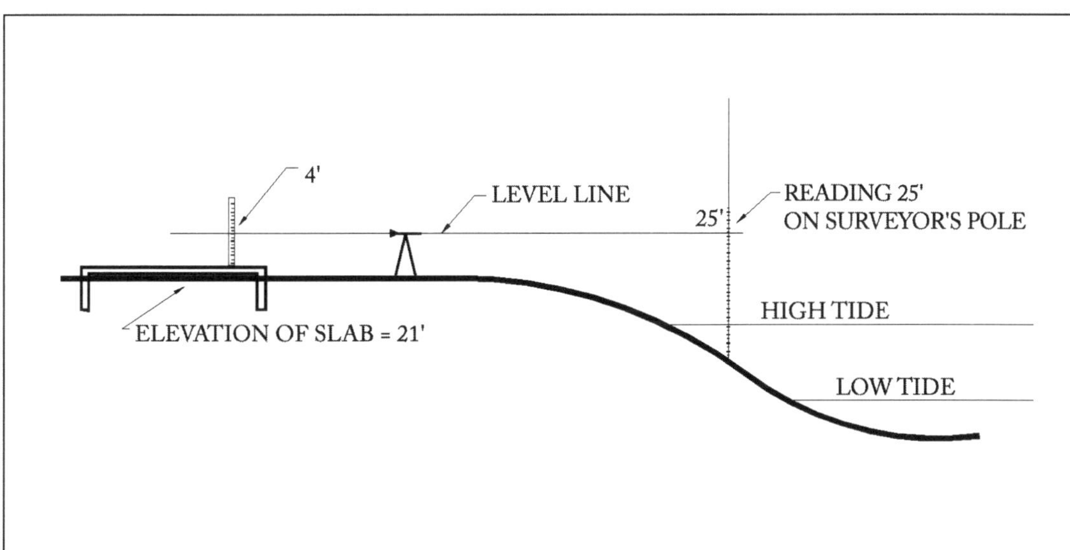

Features & Benefits

Paperwork

Feature: Homes are built according to building codes.
Benefit: Ensures the home is constructed in accordance with a governing body's regulations.

Feature: Land use is governed by the local planning and zoning board.
Benefit: Controls the growth of a town.

Feature: Completion bonds are posted by the developer.
Benefit: The bonds ensure that the municipality will finish all infrastructure and roads if the developer files for bankruptcy.

Feature: A selection sheet is used to convey the amenities the homeowner has chosen.
Benefit: The homeowner receives the home he or she wants.

Fireplaces

Feature: Engineered fireplaces are installed in the home.
Benefit: Fire refractory-lined firebox is in precise position for maximum efficiency and will emit heat after the fire dies.

Feature: Wood burning fireplaces have a damper.
Benefit: Cold air cannot enter the room when the fire is not lit.

Feature: Engineered fireplaces are UL-tested.
Benefit: Ensures safety in the home.

Feature: Engineered gas fireplaces.
Benefit: No cleanup (in most locations).

Kitchen & Bath

Feature: Cabinet interiors are laminated.
Benefit: Easy, damp-wipe cleaning.

Feature: Wall cabinets have fully adjustable shelves.
Benefit: Products to be stored in cabinets can dictate shelf placement.

Feature: Countertops are laminated.
Benefit: Attractive appearance, stain-resistant, and water-repellent.

Feature: Countertops are made from solid surfacing.
Benefit: Solid, attractive appearance, stain-resistant and easy-to-repair scratches.

Feature: Countertops are made from granite, a natural product.
Benefit: Extremely difficult to destroy, stain-resistant, and adds elegance and lasting beauty to the home.

Plumbing

Feature: Toilets use 1.6 gallons of water to flush.
Benefit: Conserves water.

Feature: Showerheads have water-saving devices.
Benefit: Conserves water.

Feature: Cultured marble tops are used in bathrooms.
Benefit: Solid, stain resistant material affords long life and is easy to maintain.

Feature: Faucets are installed with screen aerators.
Benefit: Provides an even bubble stream flow of water.

Feature: A pressure balancing diverter regulates bathroom showers.
Benefit: Prevents scalding in the shower.

Lighting

Feature: All lights are UL-rated and approved.
Benefit: Fixtures have been tested and pass required electrical codes.

Feature: Interior fixtures are lacquer-coated.
Benefit: It is easier to clean with a damp cloth.

8

CHAPTER

Stages of Construction

Any investment involving a substantial amount of money is usually prone to a variety of questions. The purchase of a home—a major investment—is no different. Questions are posed prior to committing to the purchase, followed by more questions as the construction of the home progresses. Because the features and benefits of a home interest the customer, they are effective selling tools for the sales professional.

Listed below are frequently asked questions and answers involving the different stages of construction of the home, along with various features and benefits that pertain to some answers provided. A section concerning the home in general, as opposed to a particular construction phase, concludes this chapter.

Sitework

What is a sediment pond?

A sediment pond is used to trap silt during construction. Every effort must be made to prevent sediment (small particles or fine particles of dirt) from leaving the site and disturbing the adjoining properties. If sediment is allowed onto adjoining land, the small particles of dirt will fill the voids in the soil and will prevent oxygen and moisture from reaching the roots of the grass. Sediment on soil has the same effect as laying plastic over the soil.

One way to avoid contaminating abutting property is to channel surface water into a special pond known as a sediment pond. As the pond fills, the heavy sediment particles sink to the bottom, allowing the clean water on top to escape through the spillway. Because the sediment that is lining the floor of the sediment pond, acts as a plastic coating, additional water entering the pond remains in the pond. This additional water does not seep into the dirt, which gives the pond a distinctly unattractive appearance. Thus, if a sediment pond looks unattractive, it is most probably functioning correctly.

Is it acceptable to build a home in a location that was once a sediment pond?

Even though testing is rarely performed, virgin ground is assumed to be 2,000 psf (pounds per square foot). If a house is constructed on bad soil (i.e., ground water, root material, old dump site, etc.), its structural integrity will be compromised. Bad soil in the ground can go undetected because footings are often not dug deep enough to detect the less than desirable soils.

However, when a sediment pond is filled, it is first cleaned of all sediment and then filled in a controlled environment with 6-inch lifts (i.e., adding 6 inches of dirt, compacting with a tamper, adding another 6 inches of dirt, compacting again, etc.). Then an engineer needs to certify that the dirt will be strong enough to support the home and is compacted to building specifications: 95 percent of 2,000 psf. This is similar to the initial mass grading of the subdivision prior to development. The areas that the dirt is used to raise the existing grade for a future road or pad site also must meet building specifications, i.e., those areas must be filled with 6 inch lifts and certified by an engineer to be within 95 percent of 2,000 psf.

What is a storm water management pond?

A storm water management pond—a retention pond or a detention pond—is usually created to maintain the water table in the area after a neighborhood is built.

Before a neighborhood is developed, approximately 80 percent of the rain water remains on the site. It is absorbed by the plants, weeds, soil, etc. The rest of the water would naturally run off the site.

When a neighborhood is built, the opposite occurs. Impervious surfaces such as roads, driveways, sidewalks, lead walks, patios, and roofs immediately direct about 80 percent of the rain water off the site, leaving about 20 percent of the rain water on the site.

When such a large percentage of rainwater leaves the site, the level of the ground water will start to drop, which reduces the depths of nearby streams and ponds. The excessive run-off will erode the off-site run-off areas. This situation is detrimental to the environment and should be avoided. Therefore, storm water management ponds, detention ponds, and retention ponds are created to retain the rain water on the site and allow it to perk into the ground, maintaining the water table levels while taking care not to destroy run-off areas.

How are storm water management ponds kept clean?

Have you ever noticed that some storm water management ponds are clean while others are dirty? In general, the clean ponds have been stocked with fish. Therefore, if you want to improve the appearance of your neighborhood's storm water management pond (not sediment pond), contact your state's Department of Environment and ask if they will stock the pond with fish. If they do not offer that service, inquire as to which fish thrive in your area, and perform the task yourself. Remember, though, only ponds that stay wet all year long should be stocked.

What is the difference between surface water and ground water?

When building a home, two different types of water are involved: surface water and ground water. Surface water is produced by rain or melted snow and accumulates on the surface of the ground. Swales are used to channel the surface water away from the home to an engineer-designated location. Ground water is the water below the ground surface that is exposed when one digs a hole. As the level of the water table drops, so does the level of streams, rivers, and lakes. Therefore, storm water management ponds, detention ponds, and retention ponds are created to hold the surface water on the site so it can perk into the ground, thus maintaining the level of the ground water table.

Why can't swales be moved?

The function of a swale, which is designed by the engineer, is to channel rainwater (or surface water) to an approved location. Notice in Figure 8.1 that the pipe between catch basin A and catch basin B is an 18-inch pipe and the pipe between catch basin B and the storm water management pond is a 36-inch pipe. This is because the 36-inch pipe is essentially carrying water from both catch basins. If, for example, in Figure 8.2, we relocate the swale in the front yard of home site 12 so the water is now channeled to catch basin A, instead of the designated catch basin B. Catch basin A has the potential to overflow because the pipe connecting both catch basins was not designed with the additional water from home site 12 in mind.

Figure 8.1 Catch basins A & B

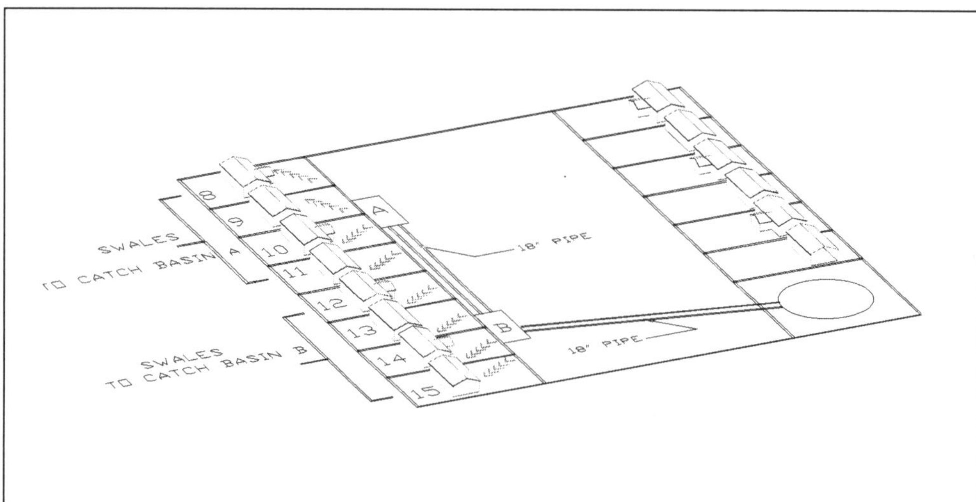

Is the sump pump location dictated by the exterior grade?

It is a misconception that the sump pump location in a basement is dictated by exterior grade conditions. Exterior grade has no influence on the location of the sump pump. Exterior grades manage surface water, whereas sump pumps manage ground water. The surface water does not go to the sump pump but is channeled by swales to a catch basin, pond, or other designated location away from the home. Sump pumps manage ground water by pumping it up and out onto the lawn, turning it into surface water.

What is the benefit of using a highlighter on a site plan?

The key to reading a site plan is knowing what the different lines represent. A great way to identify the various lines is to highlight them in different colors. For example, in order to identify topography, you have to understand the contour lines. First identify the 0 contour lines (e.g., 90, 100, 110, 120, 130…) with, say, a green highlighter. Next, identify the 2 contour lines (e.g., 92, 102, 112, 122, 132….) with a blue highlighter. Do the same with different colors for the 4s, 6s, and 8s. The end result is a color-coded plan that identifies the lay of the land. For other items such as sewer lines, water lines, easements, and LODs (Limits of Disturbances), simply highlight the different items with different colors on another site plan.

Figure 8.2 Relocated swale between catch basins A & B

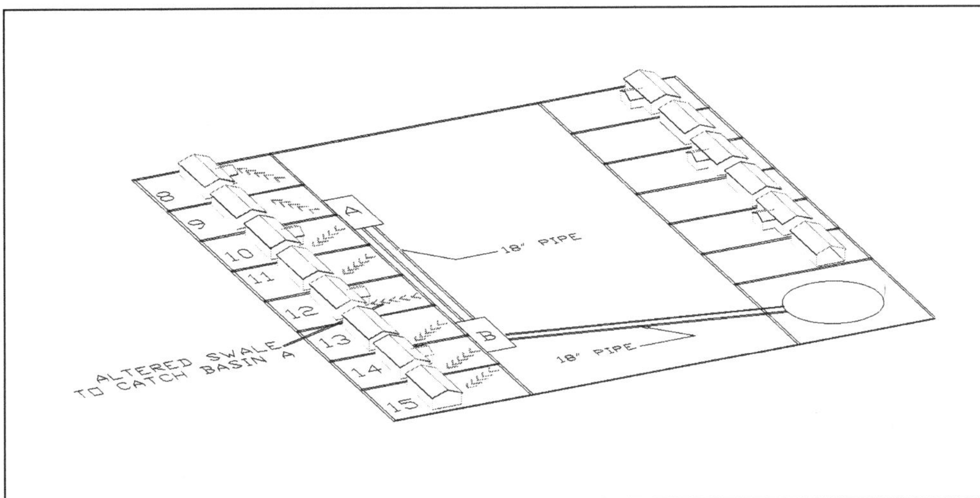

Features & Benefits

Feature: Surveyors stake out homes 10 feet wider than the actual width of the home.
Benefit: Excavation and foundation trade professionals use the same control to ensure consistency.

Feature: Holes are excavated with a 4-foot over dig.
Benefit: Minimizes settlement in the yard.

Feature: Holes are dug 4 inches deeper than the top of the footing.
Benefit: The footing will sit in undisturbed soil, and the ground water will be properly managed.

Feature: Sediment control devices such as silt fences and sediment ponds are used.
Benefit: They control soil erosion.

Foundation

Why does concrete crack?

Concrete always does four things: it cures, cracks, cures, and cracks. This is normal. As concrete cures (dries), it hydrates, which releases moisture. This causes the concrete to shrink. As it shrinks, the concrete creates pressure on itself, which can only be relieved by cracking. In fact, some concrete is designed to crack. For instance, look at your sidewalk. The straight lines you see at 3-inch to 4-inch intervals are placed there by design so that when the concrete cracks; it cracks in straight-line intervals. Typically, the less water in concrete (as measured by its slump), the less it will crack.

What is the difference between a slab and a crawl?

A slab foundation is installed directly on the ground. A crawl foundation is installed on foundation walls that sit on footings placed below the frost line. The frost line (the maximum depth at which the ground will freeze during the winter), the surrounding terrain, and the product offering typically dictate whether or not the foundation will be a slab or a crawl.

A deep frost line or a hilly terrain typically necessitate a crawl foundation because it would be difficult to place the perimeter footings of a slab at the depth necessitated by a deep frost line. With a crawl, it is easier to place the footings below the frost line, and the 4-foot foundation walls sitting on the footing will act as a retaining wall to support the exterior grade. This will channel rainwater away from the home, making grade transitions easier in hilly conditions.

Slabs are typically used in flat terrain and where the frost line is not a consideration. They are usually less expensive than crawls because you eliminate the cost of the foundation wall, floor system, and insulation necessitated in a crawl. For example, the floor of a crawl is a wooden deck that needs to be built by the framer before the walls can be built. But in a slab, the concrete is the floor on which the final flooring will be installed.

What is the difference between a mono and a two-pour slab?

Concrete is always placed, not poured. With that said, a mono slab requires placing concrete one time, making it a little easier to construct than a two-pour slab that requires placing concrete twice. Both slabs have R-8 1/4-inch rigid insulation beneath the perimeter in areas susceptible to frost. However, a two-pour slab has an extra advantage in that a thermal break can be installed. A thermal break reduces the heat loss and heat gain through the side of the slab. If frost, and hence the frost line, is a consideration, then the walls of a two-pour slab can be installed after the footing has been placed below the frost line. The concrete floor of a mono or a two-pour slab is actually the first floor of the home.

Will breaking a concrete slab in order to move plumbing pipes affect the structural integrity of the home?

Prior to placing the concrete for a slab, it is necessary to locate and install the plumbing pipes and often an electric conduit or duct work. However, if the plumber has installed the pipes in such a way that they do not fit inside the framed wall, it may become necessary to break a portion of the concrete in order to relocate the pipes a few inches. When the pipes have been moved, it is simply a matter of patching the floor. At this point, the concrete is merely supporting the final flooring.

In the case of a post tension slab, it is very important to release the tension of any cable in the proximity of the hole before breaking out the concrete. After the correction has been made, the cable is simply re-stressed. Consider the following example. Prior to placing the concrete, tub boxes are located in the slab as a form at tub and shower locations in order to create a void so the concrete does not cover the shower and tub connections. This is done to avoid breaking the concrete, making it easier to install the tub and shower traps when setting the tub or shower pan. The concept that the hole in the slab created by the tub box does not affect the foundation's structural integrity also applies to breaking a small hole in the concrete to move the pipes.

How easy is it to add a wet bar to a slab?

To add a wet bar, the slab has to be jack hammered open and a water line and drain need to be added. The complexity of adding these items after the concrete has been placed depends on the location of the nearest water and sewer lines under the slab. I would discourage adding a wet bar if it necessitates breaking through any concrete load-bearing beams that were installed as part of the slab, particularly if the beams contain rebar. Cut rebar would compromise the structural integrity of the slab. If a post-tension slab is involved, special attention must be paid to avoid hitting a stressed cable. It is always advisable to de-stress any cables in the area of the slab that will be broken. In short, a wet bar can be added after the concrete is placed, but it is not advisable.

What is the difference between rebar and welded wire mesh (wwm)?

Rebar is a structural steel rod that is placed in concrete as reinforcement and is produced in 1/8-inch increments. For example, a #3 rebar is 3/8-inch thick; a #5 rebar is 5/8 inch thick. Rebar is placed in the concrete to add tensile strength where there is a structural load. Welded wire mesh is commonly called 6x6x10 wwm, where 6 inches by 6 inches is the width of the grid pattern and 10 is the gauge of the steel. It is stored in rolls that have a rusted appearance. While wwm adds no structural value to the concrete, it does prevent the concrete from pulling apart when the concrete cracks.

Basements—Features & Benefits

Feature: Eight-inch-thick cast-in-place walls are used for in-ground foundation walls.
Benefit: Provides greater security and strength. It limits sound transmission.

Feature: Weep holes are installed at the base of the in-ground foundation walls.
Benefit: Relieves hydrostatic pressure on the exterior of the in-ground foundation wall.

Feature: The exterior of a basement wall is damp-proofed.
Benefit: Minimizes moisture wicking into the basement wall.

Feature: The soil under all concrete and around the exterior perimeters is termite treated.
Benefit: Ensures long-term protection from termites.

Feature: Drain tile and 6 inches of gravel are installed around the exterior perimeter of the basement wall footing.
Benefit: Manages water that is against the exterior of the basement wall by channeling it through weep holes to the interior drain tile system.

Feature: Drain tile is installed around the interior perimeter of the footing and drains into the sump crock.
Benefit: Forms an integral part of the ground water management system, which helps keep the basement dry.

Feature: Stone and poly are installed beneath the basement slab.
Benefit: Sealing the bottom of the slab and channeling the ground water to the interior drain tile system creates a dry and odor-free basement.

Feature: Sump pump is installed in all slabs below exterior grade elevation.
Benefit: Turns ground water into surface water by pumping the water collected by the underground drain system out to the swale.

Slabs—Features & Benefits

Feature: Batter boards are used.
Benefit: Ensures a square slab.

Feature: The soil under all concrete and around the exterior perimeters is termite-treated.
Benefit: Ensures long-term protection from termites.

Feature: Poly is installed under the concrete at all conditioned space.
Benefit: Keeps the concrete from absorbing water and becoming damp.

Feature: Post tension cables are installed in the concrete.
Benefit: Makes it extremely difficult for the slab to break.

Feature: Brick ledges are installed 1 $^{1}/_{2}$ inches lower than the top of the slab.
Benefit: Allows any moisture that has developed behind the brick wall to drain away from the interior of the home.

Feature: Point load locations are engineered into the slab.
Benefit: Ensures structural integrity.

Garage Floors—Features & Benefits

Feature: Garage floors step down at the point where the garage adjoins the home with a 4-inch curb stop.
Benefit: Stops carbon monoxide from seeping under the mudsill plate and into the home.

Feature: Garage floors slope away from the home and toward the garage door.
Benefit: Allows carbon monoxide to roll away from the home and exit through the garage door.

Feature: Garage walls abutting the home have $^{5}/_{8}$-inch fire-rated X drywall.
Benefit: Protects the home by slowing the spread of fire from the garage.

Feature: Plugs in the garage are GFCI-protected.
Benefit: Prevents accidental electric shocks.

Feature: The garage floor has a lip directly below the garage door.
Benefit: Allows the door to sit lower than the plain of the floor to prevent wind and rain from entering the garage under the door.

Framing

What is a point load?

A point load is a concentrated area of pressure created from a downward load. If it is removed the structural integrity of the home would be greatly compromised. Point loads include jack studs directly below headers, double or triple studs located directly below floor beams, and girder trusses and columns. If there is any shifting at these points, the systems above will be affected. Therefore, prior to drywall, it is important to ensure tight fits at these locations.

What do square, plumb, lined, and level mean?

I have a song that I sing in my seminars to reinforce the basics of building a home called, oddly enough, "Square, Plumb, Lined and Level." (This song, as well as others, can be downloaded as an mp3 file from BuildingWisdom.com.)

A home must be square, plumb, lined, and level to have structural integrity and quality. Every right-angle corner in the home needs to have exactly a 90-degree angle, even the slightest divergence is usually visible. The term "square" refers to 90-degree angles.

In Figure 8.3, notice that the architect has drawn the wall separating the kitchen and the dining room to be perpendicular as it intersects the back wall (i.e., the wall's inner angle is 90 degrees). Figure 8.4 shows a countertop built by the manufacturer (per the kitchen architecture drawings) at a perfect 90-degree angle. However, notice in Figure 8.5 that the framer built the intersecting wall at an 88-degree angle. The countertop will not fit exactly against the two walls. Therefore, because the kitchen corner is built out of square, it will cause the kitchen countertop backsplash to touch the wall at one end while leaving a gap at the other end. This is frequently masked by placing a piece of 1/2-inch round trim above the countertop to hide the gap.

A home's framing that slightly sits off the foundation at one corner was framed this way because the foundation was out of square. The framer had to compensate because the framer must always start and maintain "square."

Figure 8.3 Kitchen wall at a perfect 90°

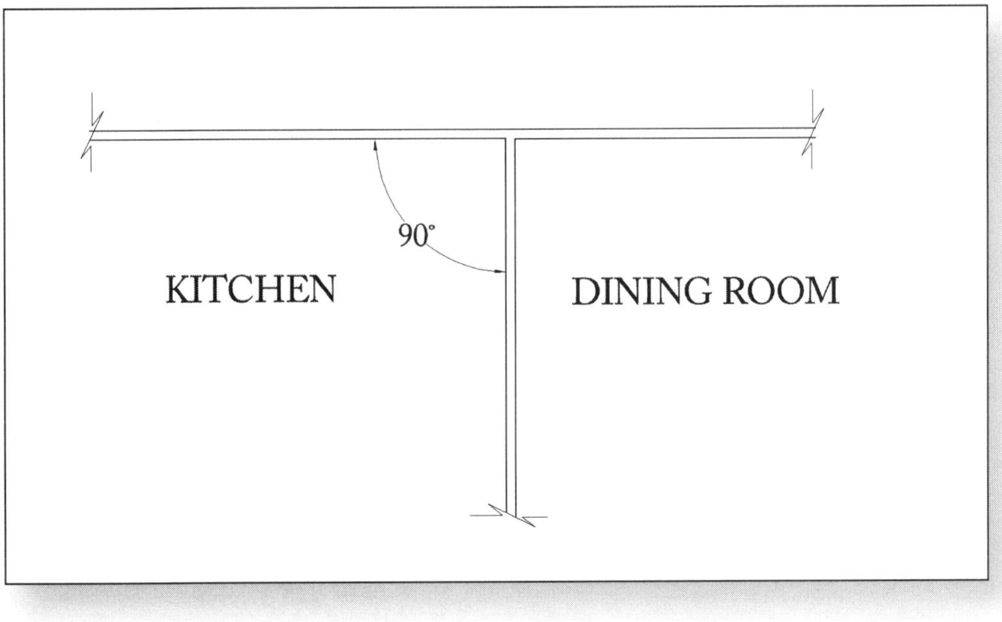

How do we ensure that everything is installed and remains square? When control is brought onto the site, the surveyor typically installs four pins, creating a large rectangle. He ensures that each corner of the rectangle is a perfect 90-degree angle. The foundation, framing, and flooring trade professionals must always begin by squaring their jobsite. To do this, the Pythagoras theorem (which we all learned back in our school years) comes into play. The foundation, framing, and flooring trade professionals use this theorem almost daily. The Pythagorean theorem states that in a right-angled triangle, a2 +b2=c2, where a and b are the lengths of the sides of the triangle and c is the length of the hypotenuse (Figure 8.6). On the job, the Pythagorean triangle's lengths are 3, 4, and 5, or multiples thereof, e.g., 6, 8, 10; 9, 12, 15; or 12, 16, 20 (see Figure 8.7).

Figure 8.4 Countertop at a perfect 90˚

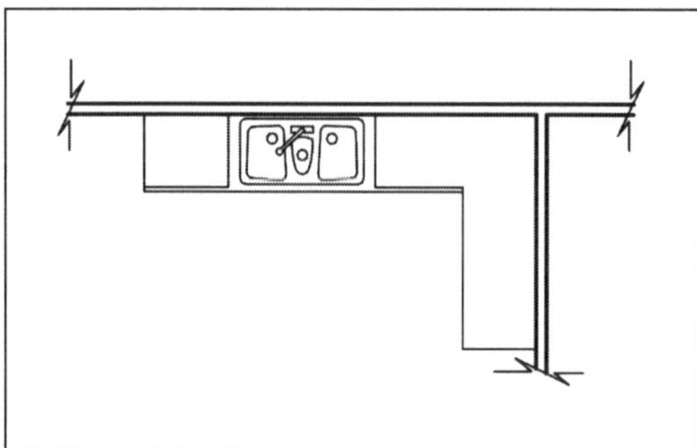

Figure 8.5 Intersecting wall not at 90˚

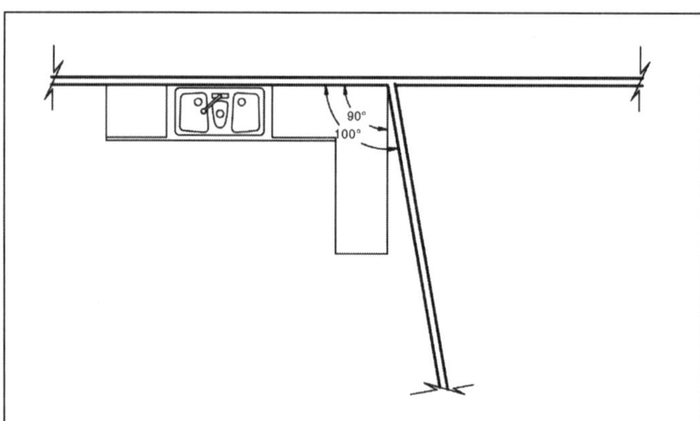

To square a building, we snap a chalk line on the foundation, representing the sidewall of the home. We will then snap a perpendicular line along the backside of the home representing the back wall (the lines intersect at a 90-degree angle). To check the accuracy of the 90-degree angle formed at the back corner of the home by these two perpendicular lines, we measure 3 inches from the corner along one line and 4 inches from the corner along the other. Then we measure the distance between the two marks. If it is 5 inches, the back corner of the home is square. If the measurement is not exactly 5 inches, this means our intersection is not square and the original lines are adjusted accordingly until the diagonal line measures 5 inches. Pulling diagonals is another way to check for square. The diagonals of a rectangle or a square should be equal. Figure 8.8 represents a rectangle out of square by 6 inches. Figure 8.9 shows the same rectangle square.

Plumb means straight up and down. If striped wallpaper looks fine throughout the room but one of the stripes disappears at a corner, then that corner is probably not plumb (i.e., not perpendicular to the floor).

Figure 8.6 A, B, C right triangle

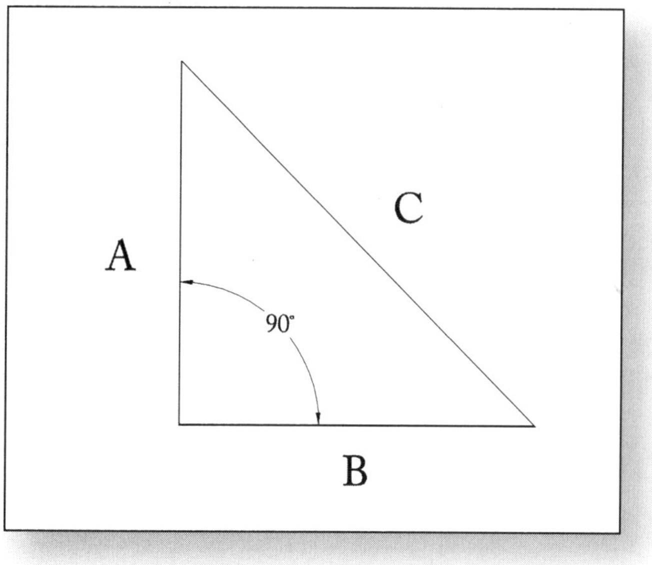

Figure 8.7 Squaring using a 3, 4, 5 right triangle

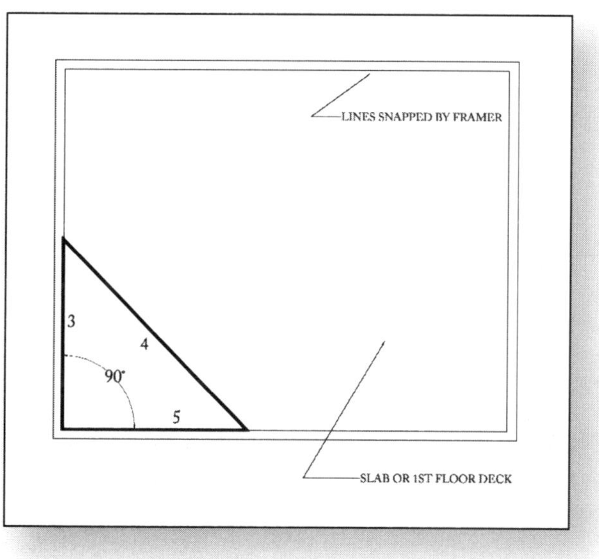

Lined means straight. An intersection that is not straight where a wall meets the ceiling is caused by the wall not being "lined" at the ceiling. The wall is lined at the bottom plate because the bottom plate follows the line the framer snapped on the slab or deck. The 2-inch-by-4-inch diagonal braces that prop up the walls as the home is built are used to line the top plate of each wall before the joist or rafters are attached. The braces must remain intact until after the sub-floor or the roof decking above have been installed.

Level means a line with no slope or one that is parallel to the horizon. A window sill, countertop, floor, etc., that slopes is said to be "out of level."

What are the differences between steel I-beams (wide flanges) and wood beams?

In a basement or crawl, the mid-bearing point for the first floor joist consists of a beam constructed from either steel or wood. Because steel is stronger than wood, the stanchion columns or concrete piers supporting the steel I-beam are typically spaced further apart than are stanchion columns or concrete piers supporting a wood beam. It also can be argued that a steel I-beam results in less floor deflection than a wood beam because wood can deflect (the reaction to compression and tension) more easily than steel.

Figure 8.8 Rectangle out of square by 6"

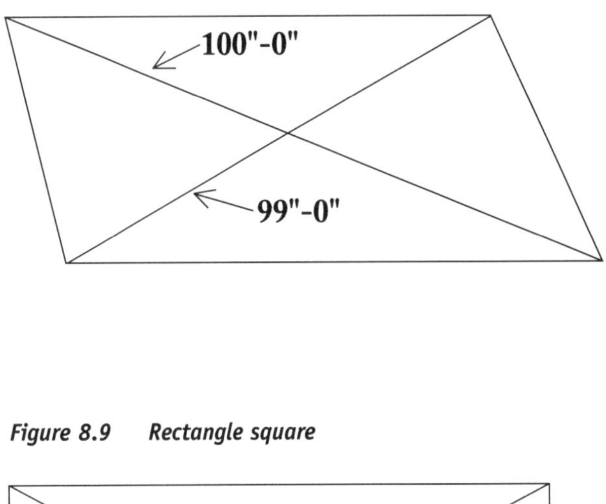

Figure 8.9 Rectangle square

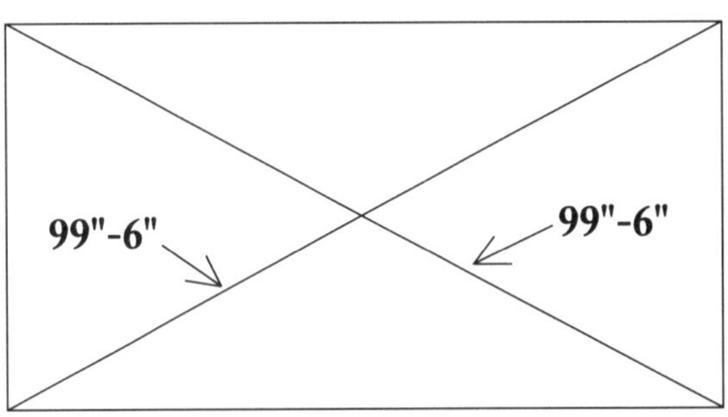

Who determines the spacing for lally (stanchion) columns?

Lally columns, also known as stanchion columns, are designed to carry the point load created by the mid-bearing beam in the basement. Their location under the beam is determined by the architect and is based upon how much weight the mid-bearing beam will need to carry and by the size of the mid-bearing beam.

If the distance between them is increased to, say, make room for a pool table, the mid-bearing beam between the columns may sag. This results in a dip in each floor and continues up through the main ridge of the home. Not only do the columns support the beam, but they may also be positioned for support at a critical point load from higher up in the building. Therefore, they should never be moved. Lally columns typically sit on a 30-inch-by-30-inch-by-15-inch (said: 30 by 30 by 15 deep) pier made of concrete, which is at the same elevation as the footing and should have been placed at the same time as the concrete for the footing.

Lally columns have a top and bottom flange, come in 3-inch increments, and are adjusted by twisting the bottom flange. This allows the framer to make any height adjustment by simply twisting the bottom bolt. Even though the bolt is strong, it should be imbedded into the concrete to ensure that the bolt does not break in the event that the lally suffers a side impact. Prior to placing the basement slab, an opening directly above the piers is set aside for their installation. The holes that are visible in the basement floor (prior to the framer's showing up) will house the lally columns.

Are bearing and spanning responsible for holding up the home?

Bearing is best visualized as one object sitting on top of another that is supporting it. Spanning is the distance between structural supports. In other words, a joist will stretch across (spans) one bearing wall to another and will be supported by its ends sitting (bearing) upon the bearing walls. Another example of a span is an OSB (oriented strand board) or a plywood sub-floor spanning the top of one joist to the top of another. The joists span the walls to create a platform, and the OSB or plywood spans the joist to create the deck.

What is the difference between an I-joist, conventional lumber, and an open web-truss floor system?

I-joists, open web trusses, and conventional lumber (2"x8" or 2"x10" or 2"x12") are designed by the architect to support the floor. The architect's seal on the plans is an indication that he or she has performed the calculations and that the spans are compatible with the lumber size and species that he or she has specified. I-joist and open web trusses have an advantage over conventional lumber in that heat ducts, pipes, or wire can be installed almost anywhere in them. With conventional lumber, it is often necessary to install heat ducts and pipes on the bottom of the joist because the code will limit where and to what extent they can be drilled. This disadvantage necessitates building a bulk head (also known as a soffit or fir down) to house the mechanicals, which will then lower the height of the ceiling at that location. Open web trusses are designed for specific distances, rendering them useless in most other situations.

What is the difference between an OSB and a plywood sub-floor?

Both OSB and plywood are designed to span from joist to joist, but each has its own characteristics. Plywood is made from large, old growth trees. Each growth ring is peeled back to become one of the layers (ply) of the plywood. Each ply is turned 90 degrees to the previous ply, giving it strength because the grain of each ply is perpendicular to the previous ply. When plywood becomes wet, it could delaminate as it dries. This may cause the plywood's span rating stamp to be inaccurate. To repair delimitation after the plywood is installed, the damaged section of plywood must be replaced with a new piece. Merely removing the delaminated plies and leveling the plywood with a floor leveling patch would result in an inadequate soft sub-floor at that location, due to the decreased number of plies spanning from joist to joist.

OSB is made from new growth trees. The grain of the wood chips from the new growth trees is similar to the layers of plywood. Unlike plywood, OSB swells when wet, but returns to its original thickness after drying. In some instances, the edges of each 4-foot-by-8-foot sheet will remain swollen after drying and will be visible through the finished floor. To correct this situation, the raised edges can be sanded to the same thickness as the rest of the sheet. Sanding will not affect the sheet's structural integrity.

Is it better to use screws or nails when attaching a sub-floor to avoid squeaks?

Let's first examine why a floor squeaks. Squeaks are produced when wood rubs against metal. Because the sub-floor is made of wood and the nails, which attach the sub-floor to the joist, are made of metal, any time that there is deflection in the sub-floor at the point of a loose nail, a squeak is produced.

So why does a nail get loose? There are three parts to a nail: the head, the shank, and the tip. The shank of the nail has rings around it. This type of nail is called a ring shank nail. When a nail is driven into a piece of wood (see Figure 8.10), the tiny fibers of the wood will fan forward, wedging the nail into place. A ring shank nail tends to stay in place better than a smooth shank nail because the wood fibers wedge themselves against each ring on the shank of the nail. But, if the nail is too close to the side edge of the joist when nailing the sub-floor, then the nail will split the joist and the fibers will be unable to hold the nail in place. Therefore, with the slightest deflection, the nail will rub against the split part of the joist, resulting in a squeak.

After installing all sub-flooring, it is always a good idea to go down to the floor below and look up at the bottom of the sub-floor. Any nails that are visible need to be backed out and a new nail installed in a different location. Gluing the sub-floor to the joist will also help prevent squeaks if done correctly. Construction adhesive should be applied only to the top of the joist that a particular sheet of sub-floor will be applied—in other words, one sheet at a time. If not, by the time the second piece of sub-floor is installed, the glue will have dried and will no longer adhere to the sub-floor, resulting in a bump in the floor. Also, as each piece of sub-floor is laid down, it needs to be nailed off 100 percent before installing the next sheet. Once again, this ensures that the glue does not skin over. Screws also help eliminate squeaks, and there is nothing wrong with adding screws to a completely nailed-off sub-floor so long as the screw does not split the top edge of the joist. In short, proper glue coverage and a ring shank nail should eliminate most squeaks. Screws, if they do not split the joist, will only help.

What is the difference between stick framing and panelizing?

Essentially, the only difference between stick framing and panelizing is that the walls of a stick-built home are framed completely on site, whereas the walls of a panelized home are framed in a factory before being shipped to the jobsite. When framing the walls on site, the framer uses the floor plan, elevations, and cross sections to correctly locate the windows, doors, point loads, etc.

Panelized walls are built on an assembly line using a specific wall panel plan drawn at the manufacturing center to ensure correct placement of the windows, doors, point loads, etc. It is rumored that panelized walls are less forgiving of mistakes caused by the foundation contractor. In any event, whether the walls are panelized or not, the mistakes in either a wall panel or in the foundation can be corrected by the framer. Because the framer is capable of framing an entire building, he or she certainly is capable of correcting any imperfections in the wall panel without much difficulty. One advantage of panelized walls over non-panelized walls is a reduction in lumber theft from the jobsite. Although all lumber is stacked at the home site, in order to steal the lumber in the wall panel it would first need to be taken apart, thus turning it into used lumber.

Figure 8.10 Fibers holding a nail in place

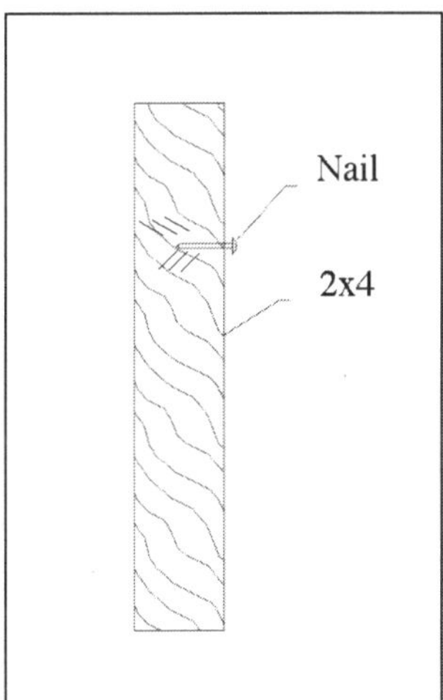

What does "racking" mean?

Racking is the tendency of a structure to become out of square (no longer 90 degrees). Homes, as well as windows and doors, should not rack. Note the framed wall in Figure 8.11. The angle where the studs meet the top is 90 degrees. In Figure 8.11, a force was applied to the top plate giving it the appearance of a parallelogram, i.e., the angle where the studs and top plate meet is no longer 90 degrees. The wall is now referred to as "racked." To prevent a home from racking, 1-inch-by-4-inch diagonal bracing can be notched into the studs from top plate to bottom plate; a diagonal metal strap can be nailed from top plate to bottom plate at both corners of the wall; or T/ply, OSB, or plywood can be installed at each corner of the home. Essentially, everything in the home should be at a 90-degree angle. It is especially important not to rack a door or window at the time of installation, because they won't close correctly and will allow air infiltration.

What is the difference between the many different wall sheathings: OSB, 1/8-inch foil sheathing, plywood, Styrofoam, etc.?

The main reason for using OSB, plywood, or $^1/_8$-inch foil sheathing as wall sheathing is to prevent the wall from racking (see Figure 7.12 to explain racking). Figure 8.13 depicts the top view of an exterior wall. Notice that the wall is 4 inches thick—3 $^1/_2$ inches for the stud wall and $^1/_2$ inch for the sheathing—at the ends of the wall but only 3 $^1/_2$ inches thick in the middle of the wall. (There is no $^1/_2$-inch sheathing applied to the middle of the wall).

If siding were nailed directly onto the wall with this $^1/_2$-inch difference in the middle of the wall, the siding would not lay straight. The result is in an indentation in the middle of the wall. To avoid this situation, the middle portion of the wall is filled with $^1/_2$ inch of either an OSB, plywood, or Styrofoam product. These pieces not only serve as filler, they also add R-value to that part of the wall. On the other hand, if 1x4s are notched into the frame at the corners or metal straps are fastened from top plate to bottom plate at the corners, then plywood, OSB or $^1/_8$-inch foil sheathing as corner bracing would be unnecessary. Additionally, a wall-sheathing product that will add a higher R-value to the wall also may be used to cover the exterior of the wall. The main thing to remember is that studs hold up the home while the sheathing, 1x4s, or metal straps prevent the walls from racking.

Figure 8.11 A racked wall

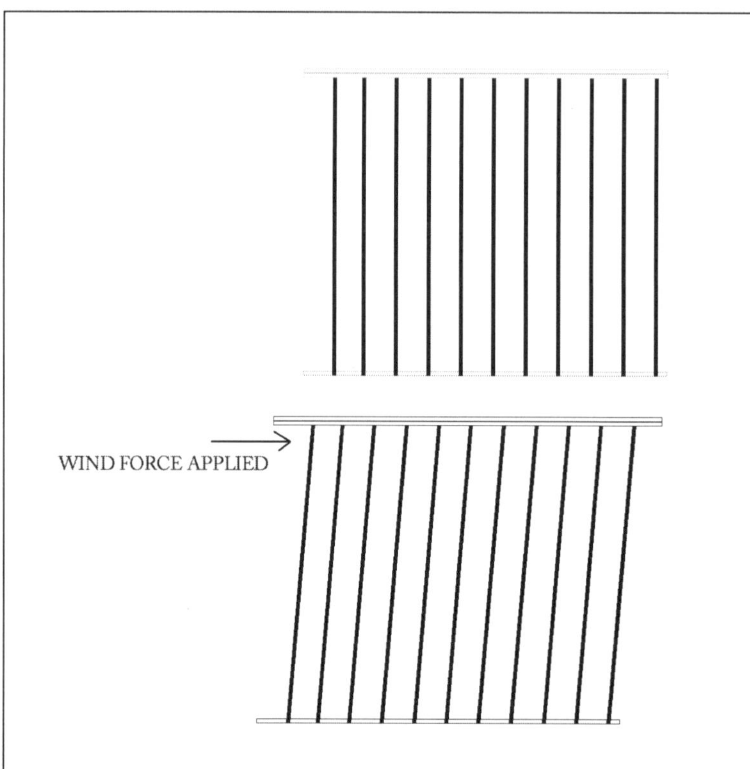

What is the difference between party walls and firewalls?

The best way to visualize a firewall is as a wall that would prevent fire from spreading to the other side of the wall or ceiling. In a condominium the ceiling where two units stack or the garage ceiling where an inhabitable space is above is a firewall. Fire walls are rated according to the time it takes to burn through the wall, e.g., a one-, two- or three-hour rated firewall. A 5/8-inch fire-rated X drywall has a 30-minute fire rating.

A party wall is a wall that separates two dwelling units and is also a firewall. Typically, codes require that two dwelling units—say units A and B—within 6 feet of each other must each have a one-hour rated fire wall. If the two walls touch each other (as in townhomes), the result would be a party wall that has a two-hour fire rating. The 5/8-inch fire-rated X drywall is installed on the living side and within the party wall of both units A and B. Because four sheets of 5/8-inch fire-rated X drywall are required to build the party wall, the party wall has a two-hour UL fire rating. A party wall may also serve to reduce noise because, if insulated, it not only reduces heat loss and heat gain but also acts as a sound barrier. Cinder block walls also can serve as party walls because they are both fire-rated and also are good sound barriers.

What is the code that dictates window placement?

For a room to be considered a room, it must have windows containing an amount of glass that is at least 8% of the square footage of the floor and its vented (openable) space is at least 4% of the square footage of the floor. An exhaust fan, despite its capacity to control odors and humidity in a bathroom, is installed as an exception to the 4% rule. A direct-wired light fixture is installed to satisfy the 8% light rule. Without the exhaust fan and direct-wired light, the bathroom would not be considered a "room."

Figure 8.12 Top view of exterior wall

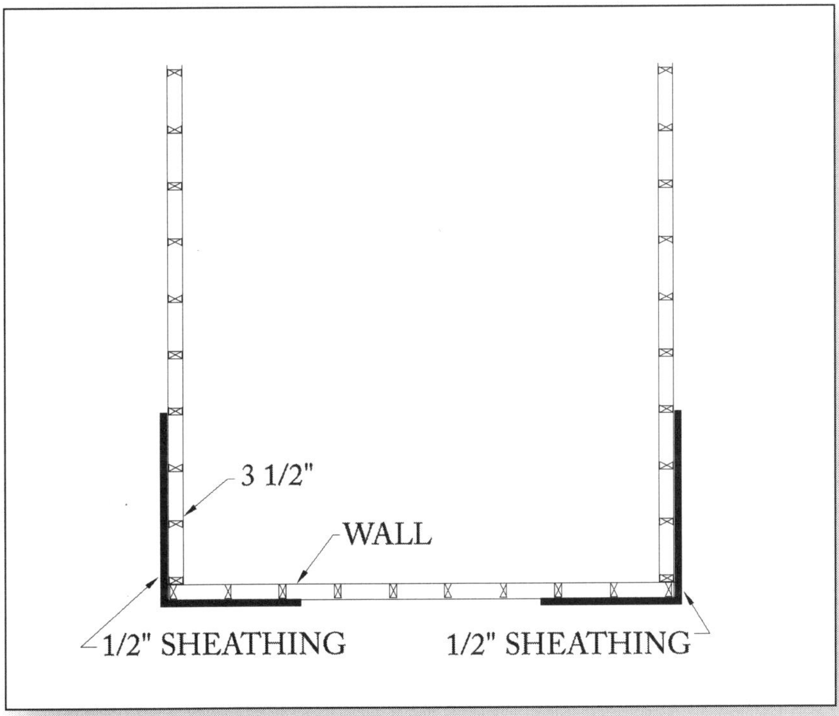

What is the maximum space between the studs in a load-bearing wall?

Studs in a load-bearing wall (i.e., a wall that supports loads from above, such as another floor or a roof) are spaced no greater than 16 inches on center, which means that the distance from the center of one stud to the center of the next will not exceed 16 inches. They also have a single bottom plate and a double top plate.

Can a load-bearing wall be moved?

A load-bearing wall is a wall that has something sitting directly on top of it. It also sits directly above an engineered location in the home that supports the weight of the load-bearing wall. If a wall is moved to enlarge a room, it is possible that the ceiling joist would then be too short to reach the moved wall, and a longer joist would need to be used. However, the current joist may be at its maximum span listed in the codebook. If open web floor trusses are used, they will need to be re-engineered because their span will change. If the wall were moved, it would no longer sit above the engineered point load location. Also, moving walls changes the size of the rooms on either side of the moved wall; therefore, special attention must be paid to the 8 percent-4 percent rule. Thus, in order to move a load-bearing wall, the home should be re-engineered.

What is the maximum space between the studs in a non-load bearing wall?

Studs in a non-load bearing wall, (i.e., a wall that is used to divide living spaces, such as creating rooms or closets), should be placed at a distance of 24 inches or less from the center of one stud to the center of the next. Non-load bearing walls are required to have only a single bottom plate and need only a single top plate.

Why can't some non-load-bearing walls be moved?

A non-load-bearing wall is nothing more than a partition wall separating a space into two smaller spaces. The wall has no structural value; its purpose is to support the drywall and define the spaces. Almost all non-load-bearing walls can be moved or eliminated without affecting the structural integrity of the home. However, the non-load-bearing walls that house heat ducts or plumbing pipes cannot be moved because either the ductwork or plumbing pipes would be exposed. Also, as with load-bearing walls, special attention must be paid to the 8 percent-4 percent rule.

What is an egress window?

All rooms in a home require one mean of egress. The door you use to enter the room is the same door used as a means of egress. A bedroom, however, requires two means of egress, i.e., the door you use to enter the bedroom and a window (egress window). The window sill must be no higher than 42" off the floor; its openable width must not be less than 20"; its openable height cannot be less than 24"; and the opening must be at least 5 sq. ft. for a first floor bedroom window and 5.7 sq. ft. for a second floor bedroom window.

Is a single-hung widow easier to install than a double hung window?

It is much easier to install a single-hung window than a double-hung window because the top sash of a single-hung window is fixed, which keeps the window square, whereas the top sash of a double-hung window is not fixed, making it very difficult to install without racking (see Figure 8.12). If the window is installed racked, which could void the warranty of the window, the sashes will not close completely, the windows may leak, and they will be difficult to lock.

Do windows from different manufacturers have different rough openings?

A 2852 (said: two eight five two and measuring 2 feet 8 inches by 5 feet 2 inches) window from one manufacturer is not necessarily the same size as a 2852 window from another manufacturer. The difference in size may be as small as a 1/2 inch, but if the framer builds the headers for the windows according to one manufacturer's framing measurements (called the rough opening) and windows from a different manufacturer are shipped, there is a good possibility that the windows will not fit into the rough openings built by the framer.

Why are man-made exterior wood products used?

Several man-made products are available for exterior use and are used by most builders to avoid potential decay such as rotting wood. Several years ago paints contained lead, which prevented mold from developing on the exterior wood. Woods with natural rot resistance, such as cedar or redwood, decay at a much slower rate than other woods. Still, they are rarely used as exterior trim because they are so expensive.

Man-made products include such items as plastic or metal for rails and columns and are fast replacing natural wood products. Man-made wood and plastic products also are replacing exterior wood trim. Sculptured pediments that are placed above the front doors or on top of windows are being replaced by man-made Styrofoam products that can be molded into a variety of shapes and sizes. Materials such as vinyl, aluminum, masonite, and fiber-cement plank siding are widely being used in place of wood siding. Because of their long-lasting, maintenance-free qualities, man-made products are more attractive to both the builder and the homeowner.

What are the different types of lumber used to build a home?

All lumber involved in building a home has been selected by the architect (per the tables in the code book) to be not only of a certain type but also of a certain size. Sizes generally include 2"x4", 2"x8", 2"x10", or 2"x12", 4"x4" and 4'x8' sheet goods. The stamp on the lumber usually includes information regarding the grade, span rating, thickness of the lumber. The stamps are often smudged, so you may have to look at several pieces of lumber before you find one that is legible.

Is it permissible not to cover the lumber package with plastic?

Let's use some common sense here. First of all, you should never wrap lumber in plastic—lumber should only be covered with plastic. By wrapping the lumber in plastic, the rain will stay off the lumber; however, without any ventilation, condensation may develop under the plastic, which could cause far more damage to the lumber than the falling and subsequent evaporation of rain on the lumber. The lumber package can be left for a week or two exposed to the elements so long as the bands that strap the lumber together are intact. If lumber becomes wet but remains banded together, it will dry straight. If the bands are not intact, any loose lumber that becomes wet may twist as it dries.

Manufactured wood stairs should never be stored in such a way as to permit water to collect where the riser, tread, and stringers meet. This could cause the lumber to swell, subsequently causing the stairs to squeak. If the steps are going to be set within a day or so from the time of delivery, they can be turned on their side to prevent water from collecting. However, if the stairs are going to be outside for any length of time, they should be stored in a dry place. Putting the stairs on skids to elevate them off the ground and covering (not wrapping) them in plastic with proper ventilation should keep them protected. Oak stairs, however, only should be stored in a dry location and never exposed to the elements. Exterior wood trim (cornice) also should be stacked and stored in a dry location.

Features & Benefits

Feature: Exterior walls are built with CMU cinder blocks.
Benefit: Protect against high winds (hurricanes) and prevent termites.

Feature: Lintels are installed on the top of all first-floor concrete walls.
Benefit: Creates a strong connection against hurricanes for second-floor joist or roof rafters.

Feature: Cells are installed in the exterior walls and next to all exterior headers.
Benefit: Keep the top lintel tied to the slab.

Feature: Angle rebar ties the top lintel, cells, and slab together.
Benefit: Ensures that the slab rebar, cell rebar, and lintel rebar are all tied together.

Feature: Steel studs are used in the home.
Benefit: Keep walls straight, prevent nail pops, and cannot be eaten by termites.

Feature: Laminated beams are installed as headers and level beams.
Benefit: Create a more spacious home.

Feature: OSB is made from new growth trees.
Benefit: Makes home building environmentally-friendly.

Feature: OSB swells when wet and returns to its normal size after drying.
Benefit: The sub-floor does not loose its structural integrity.

Feature: Sub-floor lumber is "tongue-and-groove."
Benefit: Makes the floor sturdier by resisting deflection.

Feature: Sub-floors are glued and nailed with an 8-penny ring shank nail.
Benefit: Produces a sturdier and quieter floor.

Feature: Joist hangers are used in the home and provide for continuous load transfer.
Benefit: Ensure structural integrity and provide many floor plan options.

Feature: Hurricane straps are installed on the frame.
Benefit: Protect the home from unusual weather.

Feature: Anchor bolts are embedded in the concrete foundation.
Benefit: The wood structure is securely attached to the concrete.

Feature: H-clips are installed in the roof decking.

Benefit: Transfer deflection from one sheet of roof decking to another.

Feature: Rafter ties connect the rafter (or roof trusses) to the top plate.
Benefit: Prevents the roof from blowing off the home during high wind conditions.

Feature: Fireplaces are installed with crickets behind the chase.
Benefit: Prevents rain water from leaking into the fireplace chase.

Feature: An exterior wrap is installed on the home.
Benefit: Creates a moisture barrier and helps prevent air infiltration.

Feature: $^{1}/_{8}$-inch foil sheathing is used on the exterior of the home.
Benefit: Prevents the wall from racking, is cost effective, controls air infiltration, and is a moisture barrier.

Feature: Plywood is used on the corners of all exterior walls.
Benefit: Prevents the wall from racking.

Feature: 1x4 corner bracing is notched into the top plate, studs, and bottom plate of all exterior corners.
Benefit: Stops the wall from racking.

Feature: Metal strapping is installed at all exterior corners.
Benefit: Prevents the wall from racking.

Feature: Celotex is used on exterior walls.
Benefit: Protects the wall from condensation and adds insulation.

Feature: Exterior doors are equipped with an adjustable threshold.
Benefit: Ensures a year-round airtight fit at the bottom of the door.

Mechanicals

How are air conditioning and heating units sized?

All air conditioning and forced air heating units are sized according to the results calculated from the "Manual J." A Manual J takes into consideration such things as the total volume of cubic footage of air to be conditioned, the geographic orientation (North-South-East-West) of the home, and the R-values of the exterior surfaces that border the conditioned space, i.e., the number of windows, their placement to the northern or southern exposures, and the R-value of the walls, windows, and doors. Remember, all conditioned spaces have specific characteristics. Merely assuming that a 3-ton unit would suffice in any situation is wrong.

Why should doors to rooms stay open to properly heat and cool the rooms?

The return air grills for a forced hot air and air conditioning system pull the air out of the room and send it to the air handler so it can be conditioned (heated or cooled). The registers in the room return the conditioned air to the room. If a particular room does not have a return air grill, then the air must leave that room in order to reach a return air grill. If the door to the room is closed, the air in that room cannot reach the return air grill, and the process of conditioning is severely compromised. Even though there should be a gap at the bottom of door to allow a small amount of air to leave the room, if the door is kept closed for an extended period of time, the room will probably not stay conditioned.

What is the difference between copper and CPVC water piping?

By their very names, copper and CPVC piping have a different composition. Slabs notwithstanding, my preference is CPVC piping, which is more homeowner-friendly than copper piping. CPVC piping's interior is not susceptible to mineral build-up. To repair or add to a piece of copper pipe, the system is first drained and the repair area is made perfectly dry before a blowtorch can solder the connections. Unfortunately, copper pipe is often located inside a wall and extreme caution must be taken when using a blowtorch. To repair a CPVC pipe, the system must still be drained, but the inside of the pipe does not need to be completely dry. A small amount of glue is all that is required to seal the connection. Additionally, the mineral build-up inside copper pipes may, over a period of time, cause the interior of the pipe to become narrow and restrict the flow of water throughout the home.

Why do some homes have a hung sewer?

The optimal location for the main sewer line to enter a home is beneath the basement slab to facilitate installation of a basement bathroom. A hung sewer means that the sewer line (or septic line) enters the home through the wall of the basement rather than beneath the basement slab and is then "hung" from the bottom of the first floor joist. In order to install a bathroom in a basement with a hung sewer, an ejector pit must be installed because the sewer line is higher than the basement toilet, sink, and shower, and gravity will no longer drain the water into the sewer. The elevation of a sewer line is determined by the elevation of the main sewer line in the street or the elevation of the septic system in the yard. If either system is not very deep, in all probability, the home will have a hung sewer.

How can you tell how much electric service comes to the home?

In the same way that each breaker in a service panel box displays on the tip of the breaker how many amps are on that circuit (e.g., 15, 20, 30, etc.), the main breaker (the large breaker centered at the top of the panel box) also has a number on it (e.g., 120, 150, 200, etc.). The number identifies the amount of electric service coming into the home. If the main breaker is labeled 200, then the home has a 200-amp service.

Is a home GFCI protected?

Any plug within 60 inches of a wet area (e.g., sink, commode, tub, shower, wet bar, exterior, garage, or basement) must be Ground Fault Circuit Interrupter (GFCI) protected. If a current change greater than 0.05 to 0.07 amps is detected in the circuit, the GFCI device, which is either a plug or a circuit breaker, will detect the change and will trip or break the circuit in $^{1}/_{40th}$ of a second. Because a current of less than 0.10 amps will not harm humans, the GFCI tripping at 0.05 to 0.07 amps keeps us protected from electric shocks.

Why can't a refrigerator be placed on a GFCI circuit?

A common misconception is that by placing a refrigerator on a 20-amp GFCI circuit, the refrigerator's compressor would overload and thus trip the circuit. A refrigerator will not trip a 20-amp GFCI circuit unless the refrigerator malfunctions and causes the circuit to trip. In which case, the refrigerator needs to be serviced. The plugs on the 20-amp GFCI circuit include those from the garage, exterior, basement, and bathroom. The reason a refrigerator should not be plugged into a GFCI circuit is because there is a possibility that one of the other plugs on the circuit may trip the circuit when a short of 0.05 to 0.07 amps is detected and thus prevent the supply of electricity to the refrigerator until the circuit is reset. Remember, it is not the refrigerator that trips the circuit; it is another plug on the circuit that trips the GFCI circuit.

Features & Benefits

Feature: Most sewer lines are installed under the basement slab.
Benefit: Facilitates the addition of a bathroom.

Feature: All kitchens have two 20-amp circuits.
Benefit: Supplies a sufficient amount of electricity to run small appliances.

Feature: All dining rooms have one 20-amp circuit.
Benefit: Supplies a sufficient amount of electricity to run an electric carving knife or hot plate.

Feature: Whirlpool tubs are on their own GFCI circuit.
Benefit: Prevents electric shocks.

Feature: Bathroom plugs are GFCI-protected.
Benefit: Prevents electric shocks caused by water in the bathroom.

Feature: Any plug in the garage, basement, exterior, or within 60 inches of any sink, is GFCI protected.
Benefit: Prevents electric shocks.

Feature: All countertops have an outlet.
Benefit: Prevents electric cords from draping from one countertop to another.

Feature: Countertops have outlets at 4-inch intervals.
Benefit: Allows for placement of small appliances with 2-foot long cords.

Feature: Plugs are located at 12-inch intervals along walls and within 6 inches of doorways.
Benefit: Facilitates furniture placement. Lamps and TV cords are 6 feet.

Feature: Recessed lights have an automatic shut-off switch.
Benefit: Prevents fires by preventing the lights from over-heating.

Feature: Lead-free solder is used on all copper pipes.
Benefit: Reduces the risk of lead poisoning.

Feature: CPVC is used for water pipes.
Benefit: Eliminates mineral build-up inside the pipes; pipes are easy to repair.

Close in

Why are windows and holes in the studs caulked just before drywall?

In most municipalities, code requires that penetrations from one floor to the next must be filled. Special attention is given to chases, bulkheads, fir downs, dropped ceilings, and top and bottom plates. In a wall, you can find a top plate, bottom plate, and vertical studs. When the drywall is nailed to both sides of the wall, a stud cavity is formed. If a fire starts in that cavity, the smoke in the cavity would suffocate the fire, as it has no place to go. However, if the electrician drills a hole in the top plate for his wire, and the wire did not completely fill the hole, then the smoke could exit the stud cavity and allow a fire to spread. Whether the material used to fill the hole is fire-rated or not determines whether it is known as draft stopping or fire stopping.

The windows may be caulked for several reasons. The flanges on the exterior of the windows are sometimes caulked to the frame to prevent water leaks. The jambs on the inside of the windows are caulked, foamed, or stuffed with insulation to prevent air infiltration.

What is R-value?

R-value is a material's resistance factor to heat loss and heat gain through its two surfaces (inside and outside) of the material, for example the insulation, window, or door. So how much insulation does the home require? In the example in Figure 8.13, Chicago would need the greater R-value in its insulation, doors, and windows, as its heat loss/heat gain coefficient number is the greatest. Miami would be next, and Honolulu would need virtually none.

Will increasing the R-value of the exterior walls to R-19 save energy?

Save energy? Yes. Save money? Maybe not. Because 80 percent of all heat loss and heat gain occurs through the windows, increasing the R-value in the walls is not the best way to save energy. The exterior walls of a home have insulation with an R-value of 13. R-11, if you are in the southern United States. The average double-pane window has an R-value of 1.5 to 1.9. Therefore, increasing the R-value of the windows by upgrading rather than increasing the R-value of the walls will reduce heating and cooling needs. This will conserve energy and reduce energy bills.

Figure 8.13 Insulation chart

	Miami			Chicago			Houston		
	Conditioned -	Average Temperature =	Total	Conditioned -	Average Temperature =	Total	Conditioned -	Average Temperature =	Total
Winter	70	80	10	70	20	50	70	70	0
Summer	70	90	20	70	85	15	70	70	0

Why don't homes in very hot climates need insulated windows?

In some areas of the South that experience very warm temperatures, it is often more practical not to install double-pane windows. The windows' double panes can create a greenhouse effect, which causes the temperature between the two panes of glass to rise and break the seals, rendering them inefficient. Also, because the heat coefficient of the home in the South in Figure 8.13 is 20, the expense for double-pane windows becomes questionable.

What are the different types of drywall?

A home typically has three types of drywall: 1/2-inch regular board, 1/2-inch green board, and 5/8-inch fire-rated X. The 1/2-inch drywall is installed in most places throughout the home. The 1/2-inch green board is a structural board and must be used wherever something will be affixed to the drywall, such as wall tile. Concrete wonder board is an alternative to green board and the board of choice for most tile setters. If an interior fire suppression sprinkler system is not present, then 5/8-inch fire-rated X drywall is used to create a one-hour firewall where the garage abuts the home. It is often used to create a half-hour firewall on the underside of staircases that have a closet door.

Why do we use green board?

Wall tile must be glued to drywall that is structural, because it supports the tile. Regular 1/2-inch drywall is not considered structural. Green board is a structural drywall, as is wonder board.

Why do nail pops occur?

Nail pops occur when lumber acclimates, or dries out, during the first year following completion of the home. Most lumber delivered to the site comes from the mill with a relatively high percentage of moisture content, causing it to be slightly swollen. As the lumber dries during the first year of the home (mostly as a result of the dry heat of the home's forced hot air system), the lumber shrinks. As it shrinks, it pulls the attached drywall with it. However, the nails holding the drywall in place do not move. Hence, the head of the nail protrudes over the edge of the drywall, making it appear as though the nail has "popped out." To correct this situation, take a hammer and drive the nail further into the wall, dimpling the drywall but not breaking the paper. Spread some drywall compound over the nail hole and, when dry, sand and paint to match. In order to avoid all of this, nail pops can be greatly reduced by using kiln dried lumber or a central humidifier.

Why do settlement cracks occur?

As an ex-framer, I always have a chuckle when I hear people justifying settlement cracks to a customer by stating that cracks are to be expected in a new home, and that it is best to wait a few months after the home has settled before patching the cracks. This is a fallacy.

Lumber swells when its moisture content exceeds roughly 20 percent. But when it shrinks, it tends to shrink in thickness, not in length. This causes nail pops, which are typical in new home construction, to appear. Have you ever noticed that settlement cracks usually appear at the headers above windows and doors or at major point load locations throughout the home? Settlement cracks are caused by a framer's imperfect cuts at point load locations (see Figure 8.14). This causes the lumber from above to move downward and crack the drywall.

Ideally, the way to eliminate settlement cracks is to have the framer make perfect 90-degree cuts so that the lumber fits together tightly. It is, however, unrealistic to believe that every cut will be perfect. To prevent settlement cracks, the framer should fill any void created by imperfect cuts with a shim (a thin piece of wood that tapers to a fine point) before drywall. This will prevent the lumber from dropping and the drywall from cracking.

Why do stair and crown moldings crack?

Caulk is applied to areas where dissimilar painted materials touch one another. Because they expand and contract at different rates, these areas must be periodically re-caulked by the homeowner to maintain their appearance. On a staircase, the stair stringers (a thick wood product), drywall (a gypsum product), and trim (a thin piece of wood) all touch one another and are, therefore, caulked together. Because materials expand and contract due to the seasonal moisture change in the air, the caulked areas will need to be re-caulked and painted approximately every 12 months. This concept is especially important in crown molding because heat rises and makes the crown molding more susceptible to expansion and contraction.

Figure 8.14 Mis-cut at a point load location

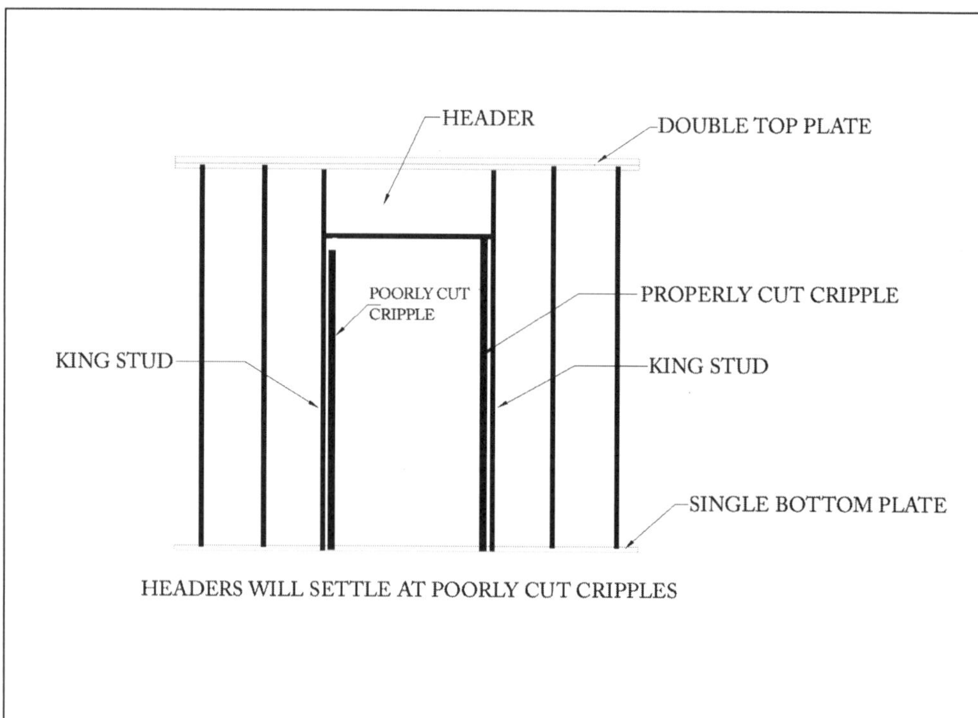

Features & Benefits

Feature: Corner bead is installed on all interior drywall corners.
Benefit: Ensures corners are straight and protected.

Feature: Batt insulation is installed so that the vapor barrier (kraft facing) faces the conditioned space.
Benefit: Creates a moisture barrier to protect the drywall from becoming damp.

Exterior

What are the differences between wood, vinyl, and aluminum windows?

A wood window has a much beefier profile, which typically reduces the area of glass and often precludes its use as an egress window without increasing its size. Wood windows are usually more expensive to install than metal or vinyl windows. Wood window installation includes the interior trim and paint or stain and requires more homeowner maintenance, unless they have a vinyl or metal clad. On the other hand, wood windows have a more traditional and authentic appearance than other types of windows; can have true divided lights as individual panes, grills sandwiched between the insulated glass, or placed on the inside face of the glass; and are available in many lines.

A vinyl widow is less apt to dent than a metal window but if it's hit it is liable to crack in colder temperatures. Installing a vinyl window is slightly more difficult than other types of windows because its frame is somewhat more flimsy. Vinyl windows are a little more expensive than metal windows but less expensive than wood windows. Because vinyl windows are slightly smaller than metal windows, egress may be an issue in bedroom windows if their size is not increased. Mutton bars (or grills) to create divided lights may be sandwiched between the insulated glass or may be placed on the inside face of the glass. Vinyl windows are basically maintenance-free.

Metal windows tend to be less energy efficient as they transfer heat loss and heat gain through the metal part of the frame and sash, and also are apt to dent if hit. They have a narrow profile that is similar to a vinyl window and can have either divided lights as muttons sandwiched between the insulated glass or placed on the inside face of the glass. Metal windows have a larger glass area than wood and vinyl windows. The larger glass area allows them to serve as egress windows without having to increase their size. Metal windows are usually the least expensive type of window and are basically maintenance-free.

What are the differences between aluminum, vinyl, masonite, and fiber-cement siding?

Aluminum and vinyl siding are both virtually maintenance-free. Because vinyl siding maintains its color throughout, scratches are barely visible. Aluminum is painted only on the exterior, so scratches usually remove the paint and make the scratch more noticeable. Aluminum siding also is susceptible to dents, whereas vinyl siding cracks. Masonite and fiber-cement siding both need to be painted. Masonite may fray if scratched or where nailed. Fiber-cement is a concrete-based product, which makes it very durable.

What is the difference between a concrete and an asphalt driveway?

Because asphalt is petroleum based, a concrete driveway will better withstand gasoline, oil, or diesel spills. In the winter, however, the top of a concrete driveway can be damaged by salt picked up by cars from the road. Once installed, a concrete driveway is virtually maintenance-free. Many asphalt driveways are sealed to enhance their appearance, but once sealed, the driveway is no longer maintenance-free. Most asphalt contractors agree that it is a better idea to allow an asphalt driveway to weather naturally.

Why do brick pavers look different than the brick on the wall?

Brick pavers on the walkway have a different composition than the brick on the wall of a home. Pavers are usually a solid structural brick with a different texture, size, and weight than veneer wall brick.

Like other colored products, bricks have dye lots. Even though the brick manufacturer made every effort to produce an identical batch, a batch of bricks produced on a given day will be slightly different in composition and color compared to a batch produced at a later date. This is because the ingredients for the two batches of brick came from different locations in the mine where the color and composition of the earth was slightly different. The dye lots of the brick paver and the matching wall brick will be different.

Features & Benefits

Feature: Exterior brick walls are installed with weep holes at 4-inch intervals in the first course.
Benefit: Allows proper ventilation, which prevents condensation.

Feature: Seamless gutters are aluminum and custom-made for each home.
Benefit: Prevents leaking.

Feature: Like-wood is used for decorative trim.
Benefit: It is a man-made, rot-resistant product.

Feature: Shingles are thermally activated and will seal together when exposed to warm sunlight.
Benefit: Roof becomes a tighter blanket and is more resistant to the elements.

Feature: Ridge vents, soffit vents, and baffles are an integral part the roof system.
Benefit: Reduces attic heat in the summer and condensation in the winter.

Feature: Gable vents, soffit vents, and baffles are used in all roof systems.
Benefit: Reduces attic heat in the summer and condensation in the winter.

Feature: Air hawks, soffit vents, and baffles are used in all roof systems.
Benefit: Reduces attic heat in the summer and condensation in the winter.

Feature: Clay roof tiles are used on the roof.
Benefit: Very resistant to wind, water, and uplift.

Feature: Exterior concrete is broom-finished.
Benefit: Ensures a safe walkway during bad weather.

Feature: Swales in the yard start 10 feet from the home and are a minimum of 6 inches lower than the home.
Benefit: Storm water is channeled away from the home.

Feature: Storm water management ponds, retention ponds, and detention ponds are created.
Benefit: Surface water is collected, thereby ensuring harmony with nature by protecting ground water levels.

Feature: Exterior concrete is air-entrained.
Benefit: Helps prevent concrete from cracking in areas that have frost.

A

APPENDIX

Cross-Section Diagram Identifying Parts of the Home

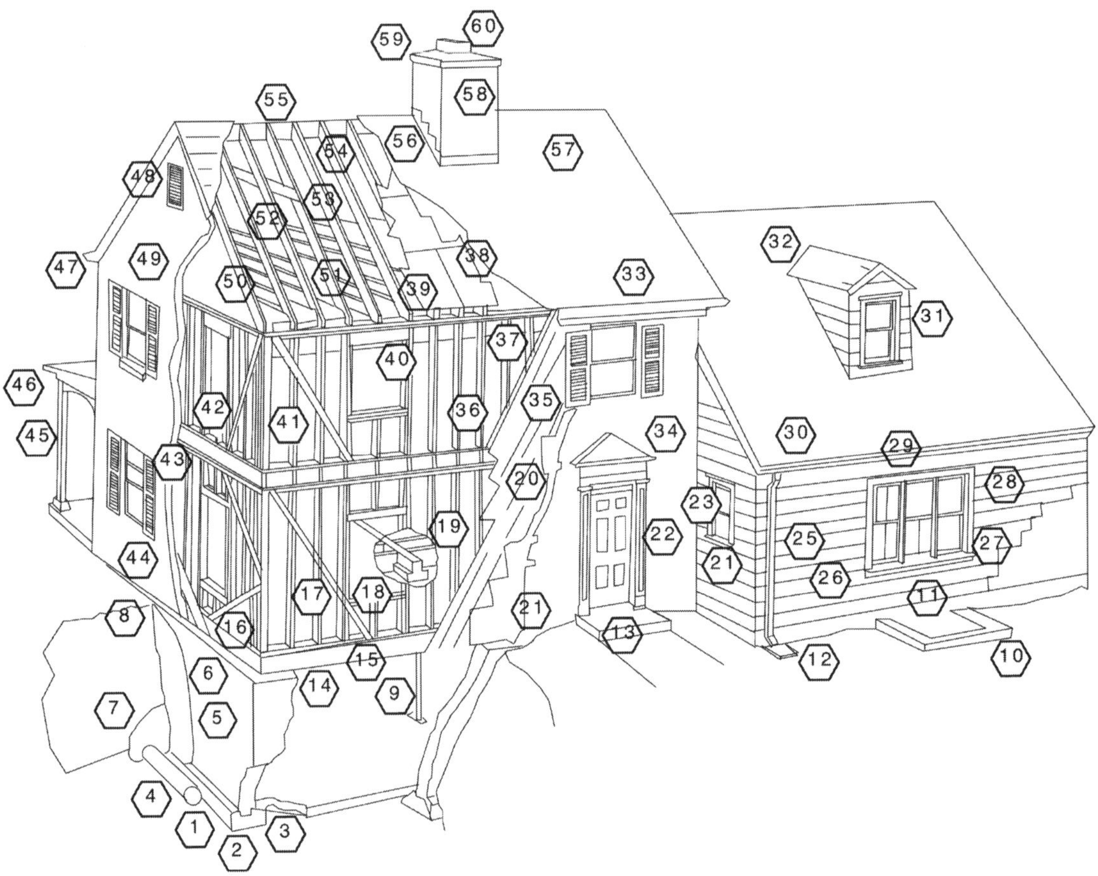

1. Footing
2. Reinforcing Rod
3. Keyway
4. Drain Tile
5. Foundation Wall
6. Waterproofing
7. Gravel Fill
8. Grade Line
9. Metal Column
10. Areaway Wall
11. Basement Window
12. Splash Block
13. Stoop
14. Sill Plate
15. Corner Brace
16. Knee Brace
17. Bridging
18. Floor Joist
19. Beam Girder
20. Sheathing
21. Building Paper
22. Trim Pilaster
23. Double-hung Window
24. Window Sill
25. Downspout
26. Bevel Siding
27. Fiberboard Sheathing
28. Window Trim
29. Mullion
30. Rake Mold
31. Dormer
32. Valley
33. Gutter
34. Shutter
35. Pediment Door Trim
36. Finish Flooring
37. Stud
38. Roof Decking
39. Double Top Plate
40. Flooring Paper
41. Corner Post
42. Subfloor
43. Lintel
44. Brick Sill
45. Porch Post
46. Porch Frieze-beam
47. Return Cornice
48. Louver
49. Brick Veneer
50. End Rafter
51. Insulation
52. Ceiling Joist
53. Collar Beam
54. Common Rafter
55. Ridge Board
56. Flashing
57. Shingles
58. Chimney
59. Cement Wash Cap
60. Chimney Flues

B

APPENDIX

Sales Center Plan Checklist

You'll find out that it's a good idea to have the following plans in your sales center.

Plan Type	Last Approval Date	Revision Date
1) Site Plan	__________	__________
2) Sediment Control Plan	__________	__________
3) Recorded Plat	__________	__________
4) Landscape Plan	__________	__________
5) Power Plan	__________	__________
6) House Plan	__________	__________

C

APPENDIX

Product Checklist

Complete the Product List form by (i) writing the vendor's name and contact information beside each product, (ii) contacting each vendor for product brochures and a list of features and benefits, and (iii) organizing the information in a 3-ring binder. Remember, the product vendors' sales professionals are usually willing to provide you with the requested information because you are helping sell their products.

Product	Vendor	Phone	E-mail	Web site
Appliances				
Bath Accessories				
Brick				
Cabinets				
Carpet				
Doors – Exterior				
Doors – Interior				
Door Knobs				
Engineered Fireplace				
Faucets				
Hardwood Floors				
HVAC Equipment				
Insulation				
Interior Rails				
Jacuzzi Tubs				
Lighting Products				
Resilient Flooring				
Sheathing				
Shelving				
Shingles				
Shower Doors				
Siding				
Sliding Glass Door				

Product	Vendor	Phone	E-mail	Web site
Tile				
Toilets				
Tubs				
Tops – Kitchen Counter				
Tops – Vanity				
Trim – Interior				
Trim – Exterior				
Windows				

D

APPENDIX

Shopping the Competition Checklist

Knowing both your company's and your competitors' construction processes will help set you apart from your competition. Walk your projects and your competitions' projects with your supervisor, and complete the following table. This drill also doubles as a great team building experience for you and your supervisor.

Task	**Your Homes**	**Competitor 1**	**Competitor 2**
Air Infiltration Package	________	________	________
Appliances	________	________	________
Cabinet Blocking	________	________	________
Cooktop Venting	________	________	________
Exterior Doors	________	________	________
Exterior Sheathing	________	________	________
Fireplace	________	________	________
Floor System	________	________	________
Foundation Type	________	________	________
Insulation			
Ceiling	________	________	________
Walls	________	________	________
Basement	________	________	________
Overhang	________	________	________
Interior Doors	________	________	________
Interior Rails	________	________	________
Locks	________	________	________
Plumbing Pipes	________	________	________
Roof Sheathing	________	________	________
Roof System	________	________	________
Shelving	________	________	________
Sub-Floor Sheathing	________	________	________
Use of Sump Pump	________	________	________
Water Heater Size	________	________	________
Windows	________	________	________

E

APPENDIX

Sales Center Inventory Checklist

- ❒ Construction File
- ❒ 3-Ring Binder
- ❒ Boots
- ❒ Hard Hats
- ❒ Paper Booties
- ❒ 100-Foot Nylon, Two-Sided Tape Measure
- ❒ Large Golf Umbrella
- ❒ *Tucker's Toolbox*
- ❒ Cover Sheets
- ❒ Site Plan
- ❒ Recorded Plat
- ❒ Sediment Control Plan
- ❒ Landscape Plan
- ❒ Power Plan
- ❒ Architectural Plans
- ❒ HOA Documents

Glossary

Acoustical Tile — A type of tile for walls and ceilings composed of mineral, cork, or metal.

Aerator — A removable extension at the end of a sink faucet containing air ports and a screen. It mixes water with air in order to prevent splashing and save water.

Aggregate — Rocks used as a fill in concrete.

Air Duct — Pipes which transport cold and warm air from rooms in the house to the air conditioning system or furnace, or vice versa.

Air Gap — A safety device installed on kitchen sinks that prevents dirty sink water from being drained back into the dishwasher.

Air Hammer — A banging noise heard in plumbing pipes, which is caused by air trapped in the pipes.

Ampere — The measurement of electricity.

Anchor Bolt — A bolt used to secure the sill plate to the foundation wall.

Anchor Strap — A strap used to secure the walls to the foundation.

Apron — A paved or concrete area adjoining the street or garage entrance.

Areaway Drain — A drain system for the basement door, and which is below the grade of the slab.

Ash Dump — A container under a fireplace where ashes are temporarily deposited prior to removing through a cleanout door.

Asphalt — A material similar to tar, used to pave driveways.

Asphalt Sealer — A substance composed primarily of petroleum, used to seal and protect the surface of asphalt from water.

Attic The space between a home's roof rafters and top floor ceiling beams.

Backfill The earth or gravel used to fill in the space around a building wall after the foundation is completed.

Balcony A floor system that extends from a structure and is supported by that structure.

Ball Cock A mechanism used in flush toilets that is composed of a valve connected by means of a lever to a floating ball. When the ball is raised, the valve closed; when the ball is lowered, the valve is opened.

Baluster One of a string of small poles used to support the handrail of a stairway.

Balustrade A row of balusters forming the edge of a balcony or staircase.

Band Board The floor system's outermost joist that is positioned on the sill plate and that extends around the perimeter of the home.

Base Shoe A molding used to cover gaps where the baseboard and the floor meet.

Base Top Molding A thin strip positioned on top of the baseboard to cover gaps between the wall and baseboard.

Baseboard A decorative board used at the junction of the wall and floor that protects the wall from scratches.

Basement A space of full-story height below the first floor of the home.

Basement Slab The floor of the basement.

Batt Blanket insulation.

Batten Small, thin strips that cover the joints between wider boards on exterior building surfaces.

Bay Window A window that extends out from the line of the wall.

Beam A support-load structural member.

Bearing A member that supports another member and serves to spread the load either vertically or horizontally.

Bearing Wall A wall that supports a floor or roof.

Berm A narrow shelf or flat area that breaks the continuity of a slope.

Bib/Bibcock A water faucet to which a hose may be attached.

Bleeder Nut A nut found on the underside of the interior shutoff valve that is used to winterize the home by unscrewing to drain water from the outside faucets.

Bleeding The process of drawing air out of water pipes. Also it is the seeping of resin or gum from lumber.

Blueprint A working plan or architectural drawing used on a construction site by subcontractors.

Box Stair A closed stair system comprised of treads, risers, and stringers.

Brace Lumber or other material nailed at an angle in order to provide rigidity.

Bridging Small wood or metal pieces placed diagonally between the floor joists in order to disburse weight on the floor, which increases the floor's load capacity.

Building Paper A waterproof, fiber-reinforced paper used in walls or roofs to keep out moisture.

Butt Joint The point where two pieces of wood or molding meet.

BX Cable A rubber-encased electric cable with a flexible steel outer covering.

Cantilever A projecting beam or joist, supported at one end only, which is used to hold up an extension of a structure.

Casement A window sash that opens on hinges at the vertical edge.

Casing A frame or trim of a window or door.

Cavity Wall A hollow wall formed by joined masonry walls, producing an insulating air space between.

Chairs Stands in concrete that support cables.

Chimney A stack of brick or other masonry extending above the surface of the roof that carries the smoke outside the home; smoke is carried inside the chimney through the flue.

Chimney Breast The horizontal projection of a chimney from the wall in which it is built.

Chimney Cap Concrete edging around the top of the chimneystack that helps protect the masonry from the elements and improves the draft in the chimney.

Chimney Flashing A strip of material placed over the junction of the chimney and the roof to divert water away from the home.

Circuit Breaker A device in the main electrical panel that opens and closes electrical circuits. It automatically shuts off an electrical circuit

	in the event of an overload. Once shut off, the circuit can be re-activated to allow normal circuit activity again.
Cistern	A tank which traps and stores rainwater.
Clapboard	A long, thin board, which is thicker along one edge, nailed to exterior siding.
Cleanout Door	An exterior door located at the base of a chimney for removal of ashes.
Collar Ties	A horizontal brace in the roof that braces opposing rafters.
Concrete	A mixture of cement, sand, aggregate, and water.
Concrete Dusting	A dust that gathers on the surface of finished concrete.
Condenser	A unit that is part of the central air conditioning system. It is located outside the home and emits air to the outside of the home.
Conduit	A metal or plastic pipe in which electrical wiring is installed.
Contour	An imaginary line on the surface of the earth connecting points of the same elevation.
Control Joint	A point where pressure is relieved or an intended crack.
Coping	Tile or brick used to cover the top of a masonry wall.
Corbel	A horizontal projection from a wall that supports an overhead structure or forms a ledge.
Corner Bead	An angled metal edging used to form an edge where drywall panels touch at the exterior edges.
Cornice	A decorative molding at the top of the exterior walls under the eaves.
Course	A horizontal row of bricks, cinder blocks, or other material.
Covenant	A private legal restriction on the use of land, recorded in the county's land records.
Crawl Space	A small, unfinished space below the first floor of a home with no basement or in the attic directly below the roof. It provides access to pipes and ducts.
Cricket	A small structure behind a chimney that directs water in both directions.
Cripples	Cut-off framing materials above and below windows.
Crown Molding	Ceiling trim or molding used at the point where the wall meets the ceiling.

Curved Stair	A stair system, which has dissimilar front and rear tread widths.
Damp Proofing	A moisture barrier, which involves the tarring of the foundation.
Damper	An adjustable valve at the top of a fireplace that regulates the flow of heated gases into the chimney.
Dead Load	Loads consisting of the weight of fixed objects such as the walls and floors.
Dead Wood	Wood added to the framing onto which sheetrock is secured.
Deck	A plywood subflooring formed by its attachment to the floor joists.
Dehumidifier	An electrical appliance, which removes the humidity or moisture from the air, often used in the basement during the summer.
Delamination	The separation of laminate material or top plies from the base to which they are attached.
Density	The number of dwelling units divided by the gross acreage being developed.
Dentil Molding	A type of exterior trim that resembles teeth.
Detention	Managing storm water runoff or sewer flows through temporary holding and controlled release.
Dethatching	The removal or loosening of grass and leaves from lawns, thereby enhancing the growth of grass by allowing it to breathe.
Door Buck	The rough frame of a door.
Door Casing	The trim around a door.
Doorstop	A device attached to the wall or floor to prevent a door from opening too far and damaging the wall.
Dormer	A projection built out from the slope of a roof. Used to house windows on the upper floor and to provide additional headroom.
Double Glazing	An insulating windowpane consisting of two panes of glass with a sealed air space between them.
Double-Hung Windows	Windows with moveable upper and lower sashes.
Double Plate	The top horizontal member of a framed wall used to tie exterior walls together and to support second floor joists and trusses.
Downspout	A vertical pipe composed of aluminum used to carry rainwater from the gutters to the ground.

Drain Tile	An underground pipe that leads to the sump crock.
Drainage	The removal of excess surface water or groundwater from land by means of surface or subsurface drains or swales.
Drip Edge	The projecting part of a cornice used to get rid of rainwater.
Drywall	Interior wall construction using gypsum or plasterboard as a finished material.
Easement	A right given by a land owner to another party for specific limited use of that land to allow passages of public utilities, access to another property, etc.
Eave	The overhang of a sloping roof that extends beyond the walls of the home.
Efflorescence	A white powder that can form on new brick, block, or stucco finishes.
Emergency Spillway	A vegetated earth channel used to safely convey flood discharges in excess of the capacity of the principal spillway.
Erosion	The wearing away of the land surface usually caused by running water.
Expansion Joint	The separation of the slab.
Face Frame	The front face of kitchen and bathroom cabinets to which the hinged doors are attached.
Face Nailing	The act of nailing through a finished surface so as to leave the flat top of the nail head exposed after nailing.
Fascia	The exterior horizontal trims surrounding the rafters and directly behind gutters and over gable trim boards.
Fibermesh	A kind of concrete reinforcement used instead of 6x6x10 wwm (welded wire mesh).
Filler Board	A wood used to fill gaps between cabinets and wall openings.
Fillers	Wood putty used to fill holes or cracks in wood prior to painting.
Final Grade	The ultimate placing of soil on a yard.
Fire Block	Wooden blocks that eliminate drafts between the floors of a dwelling.
Fire Door	A fire-resistant door.
Fire Rating	A number assigned to a wall or floor designating the amount of time it would take to burn through the wall or floor.

Firewall A wall constructed of fire-retardant material to prevent the spread of fire within a building. The firewall carries a standard rating that designates its ability to constrain fire in terms of hours.

Flashing Sheet metal or other impervious material used in roof and wall construction to protect a building from seepage of water.

Float The process of smoothing out irregularities in walls and ceilings.

Flood Plain Lowland that is subject to flooding.

Floor Joist Horizontal boards laid on edge resting on the beams that provide the main support for the floor. The subflooring is nailed directly to the joists.

Floor Opening The part of the deck where the joists or floor trusses are cut back to create a hole in the deck.

Floor Truss A pre-fabricated or manufactured floor joist.

Flue An enclosed passage, usually made of fire clay or terra cotta pipe, in a chimney or any duct or pipe through which smoke, hot air, and gases pass upwards.

Footing A concrete support under a foundation, chimney, or column designed to distribute the weight of the structure over the ground. It typically sits on solid ground and is wider than the structure being supported.

Footprint A drawing that shows the building's exact location on the lot.

Foundation The lower parts of walls, usually below ground level, upon which the structure is built.

Foundation Drain A plastic pipe placed next to the foundation footing to aid in water runoff.

Foundation Wall The masonry or concrete walls below ground level that act as the main support for the frame structure.

Framing The lumber of a house, i.e., joists, studs, rafters, and beams.

French Drain A drain system that relies on gravity.

Frieze Board A horizontal exterior molding resting directly below the cornice.

Frostline The depth of frost penetration in the soil.

Furring Thin strips of wood used to level a wall or used to attach paneling or drywall.

Fuse A short plug; located in the electric-panel box, that breaks an overloaded electrical circuit.

Gable	The triangular portion of an end wall that rises from the level top wall at either end of the home.
Gabled Louvers	A vent with louvers at the peak of the gable ends.
Gambrel Roof	A roof with two pitches that is designed to provide more space on upper floors.
Gang	The number of switches in a box.
Girder	A main member in a framed floor that supports the joists that carry the weight of a floor or partition.
Glazing	The act of fitting the glass into windows or doors.
Grade	The slope of a road or ground.
Grade Beam	The structural member in a slab.
Grade Line	The point at which the ground touches the foundation wall.
Green Board	A water-resistant sheet rock.
Green Lumber	Insufficiently dried lumber that often warps or bleeds resin.
Green Plate	The treated bottom piece of wood.
Ground Fault Circuit Interrupter (GFI)	An electrical device, typically installed in areas where moisture is present, which interrupts electrical power in the event of a weak electrical loss of ground.
Ground Works	Underground plumbing.
Grout	A mortar compound used to fill spaces between ceramic tiles.
Guardrail System	Part of a building system near the open sides of elevated spaces designed to prevent people from falling.
Gusset	A brace or bracket used to strengthen a structure.
Gutter	A channel attached to the overhand of the roof to carry water runoff to downspouts and away from the foundation.
Gyp-Crete	A product made of concrete used on the second floor to prevent fire and noise.
Handrail	A typically waist-high rail used as a handhold.
Hang	The act of nailing drywall in place.
Hard Hat	A protective helmet.
Hardwood	A type of wood with a close-nit grain from trees such as oak and maple.
Head Casing	The strip of molding positioned over a door or window frame.

Headers Double wood pieces that support joists in a floor or are placed over windows and doors to transfer the weight of the roof and floor to the studs.

Hearth The floor of the fireplace usually made of brick or decorative stone, which projects into the room and prevents the embers from causing a fire.

Heel The end of a rafter that sits on the wall plate.

Hip The outside angle formed by the meeting point of two slopes of a roof.

Hip Roof A pitched roof with sloping sides and ends.

Honeycomb An open texture found in concrete that occurs when the concrete is poured.

Hosebib The exterior faucet connection for lawn and garden hoses.

Hung Sewer A method of making a sewer connection when it must be made above the basement floor level.

Hurricane Ties Metal straps used to anchor roofs to the wall.

Impervious Not allowing infiltration.

Insulation Material that prevents heat loss and heat gain.

Invert The lowest point on the inside of a pipe.

Jacks Headers' support stud.

Jamb A vertical surface lining an opening in the wall, which has been left for a door or window.

Joint Compound A plaster-like material used in conjunction with drywall tape to join drywall sheets into a continuous panel.

Joists Horizontal, parallel beams directly supporting the boards of a floor or the laths of a ceiling.

Joist Hangers Metal brackets that connect the end of a joist to a structural piece that is at right angles to the joist.

Keeper Plate The metal plate used to ensure that a door-lock latch is kept in place.

Kick Plate A protective plate that is located at the bottom of an exterior door or other location likely to be damaged or worn.

Kiln-Dried Baked or dried brick, tile, or lumber from which moisture has been removed; the process is more effective than air drying.

King-Post Vertical framing piece in the center of a roof or the center post of a truss.

Knee Wall	A wall-like structure that is approximately knee high.
Lag Screws	Also referred to as "coach screws." Large, heavy fasteners used in heavy framing.
Lally Column	Circular, concrete-filled tube used to support beams and girders.
Landing	A platform located at the end of a flight of stairs or between two flights of stairs.
Lateral Force	A force that runs in a straight line and is parallel to a level plane.
Lath	Thin, narrow strips of wood fastened to the framing members onto which plaster, tiles, etc., are applied. Gypsum or metal lath, which are cheaper than wood and can be installed in sheets, also is used.
Leaching Bed	Tiles in the trenches that carry the treated liquid waste from the septic tank.
Ledger	A horizontal, wooden floor joist support beam that extends from the foundation to the roof and forms the walls of all floors.
Life Line	A support rope, attached to a safety belt or harness, which supports a person in the event of a fall.
Lintel	A horizontal top piece over a window or door that carries the weight of the walls above it.
Live Loads	A temporary or non-structural load.
Load Bearing Wall	A strong wall that is capable of supporting the weight of other parts of a home.
Lockset	A door lock.
Louver	Slats or fins over an opening, angled to permit ventilation but keep out rain or snow.
Mantel	The decorative facing placed around a fireplace.
Masonry	Construction made from brick, stone, tile, cement block, or other similar materials.
Mastic	A construction adhesive, typically applied from a caulking gun, used mainly when installing tiles.
Miter	The ends of any two pieces of board of corresponding form cut off at an angle and fitted together in an angular shape.
Moisture Barrier	A layer of treated metal, foil, plastic, or paper used to prevent moisture penetration into walls or floors.

Moldings	Strips of ornamental wood used around windows and doors, for base and tile molding, for chair rails, and for exterior molding. Moldings are applied to cover the junction of roof boards and outside walls.
Monument	A fixed object and point established by surveyors to determine property locations.
Mulch	A natural or artificial layer of plant residue or other materials covering the land surface to conserve moisture, hold soil in place, assist in establishing plant cover, and minimize fluctuations in temperature.
Mullion	The narrow vertical strip that separates two adjacent window sashes.
Muntin	Thin, vertical strips inside the window sash that divide window openings.
Nail Pops	Caused by the shrinking of wood and the settling of the home, nail pops are nails that have loosened themselves from a stud and pushed up the joint compound.
Newel	A post, located at the top, bottom, and landings of a staircase that supports the railing.
Newel Post	The corner post of a rail system.
Nosing	The rounded outer face of a stair tread that extends over the riser.
Open Rise	A stair system consisting of treads and stringers.
Open Space	The total area of land and/or water not improved by those living in a planned community.
Orange Peel	A kind of wall texture.
Oriented Strain Board(OSB)	A man-made roof sheathing.
Overhang	The portion of the roof that projects beyond the exterior wall.
Pad	Compressed soil.
Parging	The application of a rough coat of mortar or plaster applied over a masonry wall to protect a surface from water penetration; also used on the outsides of foundation walls to protect from water penetration.
Perc Test	A determination of the rate of percolation or seepage of water through natural soils.
Percolation	The movement of water through soil.

Pier A column, usually of steel reinforced concrete, which is evenly spaced under a structure to support its weight.

Pilaster A projection from a wall or the foundation wall used to strengthen the wall by adding support or by preventing buckling.

Pitch The angle or slope of a roof.

Planned Unit Development (PUD) A special classification authorized in some zoning ordinances where a unit of land under control of a single developer may be used for a variety of uses and densities, subject to review and approval by the local governing body.

Plate Straps Metal straps used to fasten down the wall.

Plates Horizontal pieces of wood placed on wall surfaces to form a base for supports.

Platform A raised horizontal surface of wood, stone, metal, etc., which is usually higher than the adjoining area.

Plenum An accumulation and distribution chamber for connection of ducts in an air conditioning system.

Plumb Absolutely straight up and down.

Pocket Door An interior sliding door.

Pointing Mortar placement or replacement of joints in masonry by finishing brick mortar and stone concrete masonry joints.

Ponding An accumulation of water on driveways, walkways, or lawns.

Porous Fill Gravel or sand placement beneath the basement floor which stops water from rising through the concrete and assists in water runoff.

Post and Beam Framing The use of heavy posts, rather than smaller studs, in order for large open areas to be created in a building

Post-Tension Applying tension to the slab after the concrete has been placed, thereby increasing its strength.

Pre-Tension Applying tension to the slab prior to the placement of the concrete, thereby increases its strength.

Proffer A development plan and/or written condition which, when offered by an owner and accepted by the Board of Supervisors, becomes a legally binding part of the regulations of the zoning district pertaining to the property in question.

Rabbet A groove or notch in a board that allows two sections to be combined with a flush joint.

Racking The causing of a structure, such as a wall or door, to shift so that it is out of plumb and square.

Radiant Heat Used to heat rooms, radiant heat consists of coiled electricity, hot water, or steam pipes.

Rafter One of a series of sloping beams that extends from the exterior wall to a center ridgeboard and provides the main support for the roof.

Rail Boot A metal floor bracket that serves as a support for safety rails.

Rake Rail The rail that goes up the stairs.

Rebar Steel bars that are placed in concrete as reinforcement.

Register A grill that covers the opening of a duct.

Reinforced Concrete Concrete reinforced with wire or steel rods.

Rendering A term used in perspective drawing, such as an artist's sketch.

Resilient Flooring made of vinyl, rubber, or other resilient material used in areas such as kitchens, halls, and bathrooms.

Retaining Wall Any wall that blocks or supports a bank of earth; any enclosing wall that holds back the lateral pressure of internal loads.

Retention The temporary or permanent storage of storm water to prevent it from entering the sewer system.

Reveal The end of a door that can be seen only when the door is open.

Ridge Vent A vent system in the peaks of roofs that helps create natural air ventilation.

Ridgeboard A heavy horizontal board placed on its edge at the top of the roof to which the rafters are attached.

Rip-Rap Broken rock, cobbles, or boulders placed on the earth's surface for protection against the actions of water.

Riser The vertical face of a step that supports the tread.

Roof Anchor A device that provides an anchor point for lifelines.

Roof Felt Sheets of felt or other close-knit material that are placed on top of the roof boards to waterproof the roof.

Roof Sheathing Sheets of plywood that are fastened to the top of the rafter or truss to tie the roof together and support the roofing material.

Roof Shingles Thin, wedge-shaped pieces of asbestos used as the outer covering for a roof.

Rough Grade The initial placing of soil on a yard.

Runoff The portion of precipitation that flows from a drainage area on the land surface, in open channels, or in storm water conveyance systems.

S-Hooks Hooks used to position cables in concrete.

Safety Belt A device typically worn around a person's waist and attached to a lifeline; used to prevent the person from falling.

Safety Shoe Footwear that meets or exceeds safety standards.

Sash A fixed or moveable frame around a door or window.

Scaling (concrete) The separating of the top surface from the rest of the concrete.

Scaling (paint) The peeling away of paint.

Scotia A type of concave molding.

Scuttle Hole An opening to the attic, crawl space, or plumbing pipes that permits access.

Septic Tank An underground tank used for depositing domestic wastes.

Service Door The entry door to the garage.

Setback The distance from a lot line or other reference point, within which no structure may be located.

Shakes Hand-cut wooden shingles used for roofing or siding.

Sheathing The first covering of sheets of material, which is applied across floors, rafters, or studs.

Sheet Rock Boards used to construct interior walls.

Shim A tapered strip of wood used to make structures level or plumb.

Shingles Thin pieces of material, such as wood, asbestos, slate, or fiberglass that are used as an outer covering on roofs or walls.

Shiplap Boards with overlapping edges.

Shoe Molding A thin strip of wood positioned at the point where the baseboard and floorboards meet to hide the joint. The shoe molding improves the aesthetics of the room and helps prevent drafts.

Shrinkage Cracks Cracks that form from drying too rapidly.

Siding Aluminum or boards nailed horizontally to the vertical studs to form the exposed surface of the building's outside walls.

Siding Lap	An overlap in the siding.
Silicone	An artificial compound used when manufacturing caulking and sealing compounds and lubricants.
Sill Plate	The lowest horizontal member of the frame that sits on top of the foundation wall and forms a base for the studs.
Silt	A type of soil.
Single-Glaze	A single-pane window.
Sink Hole	A hole formed when dirt settles.
Site Plan	A detailed plan, drawn to scale, showing the development of a parcel of land and containing all information required by the zoning ordinance.
Skim	The final coat of spackle prior to painting.
Slab	A flat, horizontal reinforced concrete area placed directly over the earth or on a gravel base.
Sleeper	A strip of wood placed directly over the concrete floor to support flooring.
Slope	The degree of deviation of a horizontal surface.
Soffit	A vent placed under the ceiling of a roof overhang; the visible underside of structural members.
Softwood	Wood from a cone-bearing tree that is more pliable than other woods.
Soil Stack	A vertical plumbing pipe to drain wastewater.
Spackle	A compound used to fill in small holes or other defects.
Spalling	Flaking off or chipping of concrete or other masonry material surfaces.
Spaulding Concrete	The cracking of concrete.
Spiral (staircase)	A stair system that curves upwards around a central point.
Stack (roof)	A vent.
Stack (sewer)	Cleanout.
Stain-Grade	A virtually flawless trim made with no joints or knots.
Stair Platform	The platform on which a staircase rests.
Stair Railing	A railing that is used as a handhold and runs parallel to the stringer.

Stanchion Footing	Footing that is under the stanchion post and supports a column or post.
Stanchion Post	A column, which supports the floor load and is evenly spaced under the beam.
Storm Sewer	A sewer that carries storm water and surface water, street wash and other wash water, or drainage, but does not include sewage and industrial waste.
Stringer	One of the sloping enclosed sides of a staircase that supports the treads and risers.
Stud Facing	The depth of a stud (usually 3 ½ inches).
Stud Thickness	The width of a stud's edge (usually 1 ½ inches).
Studs	In wall framing, the vertical members to which horizontal pieces are attached. Studs are placed between 16 and 24 inches apart and serve as the main support for the roof and/or second floor.
Subdivision Plat	A detailed drawing, to scale, showing the division of a parcel of land into two or more lots and containing engineering considerations and other information required by the subdivision ordinance.
Subfloor	Plywood sheets nailed directly to the floor joists and that serve as a base for the finish flooring.
Sump	A pit in the basement in which water collects before being pumped outside by the sump pump.
Sump Pump	A pump that disposes water that has collected under the home foundation in the sump.
Surface Capping	The placing of an additional layer of similar material over existing material.
Swale	An elongated depression in the land surface that conducts storm water into drainage channels.
T-Pole	A temporary supply of electricity.
Tack Strips	A wooden strip with exposed tack points that is attached to the subflooring and to which carpet is attached.
Tape	Paper used to cover the seams in drywall.
Texture	A finish for walls and ceilings.
Thermocouple	A safety device for gas powered equipment that will not allow the unit to be turned on unless the pilot light is burning.
Tie	A wood member used in the construction of a building.

Tie Wall Large wood members, typically built into a wall and backfilled with earth.

Tile Drain A pipe with open joints to collect and channel excess water.

Tile Field Drain tiles with open joints used to distribute septic tank waste over an absorption area.

Toe Kick The base of a cabinet.

Toenailing Driving nails into the base of a framing member at an angle.

Tongue-and-Groove A carpentry joint where the pieces of board are made so that the tongued edge of one board fits into the grooved end of a similar board.

Topography General term which includes characteristics of the ground surface such as plains, hills, and mountains; includes degree of relief, steepness of slopes, and other physiographic features.

Trap The curved section of a water pipe that holds water so sewage gases will not be emitted from the plumbing system into the home.

Tread The flat part of the stairs that rests on the riser, and where the foot is placed.

Tread Width The distance from the front of the tread to the back of the tread.

Trim Interior-finishing pieces, such as door and window casings and moldings.

Truss A rigid, prefabricated combination of structural members used for spanning between load-bearing walls or in the construction of floors and roofs.

Underlayment A flooring layer between the subflooring and the floor.

Vacuum Breaker A device on exterior faucets that permits water to flow out of the home.

Valley The depression at the junction of two roof slopes.

Valve Seat The part of the faucet valve assembly upon which the valve sits.

Vapor Barrier Material that prevents the penetration of moisture from rooms to the outside walls.

Variance A permit which grants a property owner relief from certain provisions of the zoning ordinance when, due to the particular physical surroundings, shape, or topographical condition of the property, compliance would result in a particular hardship or practical difficulty, therefore depriving the owner of the reasonable use of the land or building involved.

Vent A small aperture to allow the passage of air through any space in a building.

Vent Pipe A pipe that permits the escape of gases from a plumbing system.

Vertical Force A straight-line force that runs perpendicular to a level plane.

Wainscoting A material placed on a wall between the baseboard and the chair rail that is different to the other material on the wall.

Wall Sheathing A base for exterior siding that is attached to the outside face of studs.

Wall Ties Wires used to tie masonry or veneer to the framework of the home.

Washer A circular rubber or plastic piece that goes under a nut or bolt and acts as a sealing device.

Washing Soda A product used for heavy-duty cleaning of surfaces and also in drains traps to prevent other materials from accumulating.

Washouts An area where soil erosion has been caused by water.

Waste Line A pipe that carries waste from a bathtub, shower, lavatory, or any fixture or appliance to a sewer line or septic tank.

Water Table The upper surface of the free ground water in a zone of saturation.

Waterproofing A horizontal layer of water-resistant materials placed between the interior and exterior walls to prevent moisture.

Watershed The area of land that drains into a stream, lake, or other body of water.

Weather-Stripping An insulating strip of material around windows and doors that reduces water penetration or air infiltration into the home; also reduces the extent of conditioned air from escaping from the home.

Weep Hole A small hole in a wall that allows water to drain off.

Winders A wedge-shaped step that has a wider tread on the outside, thereby changing the direction of the stairs.

Window A framed opening in a wall that holds a pane of glass.

Window Balance A device in windows that aids in the opening and closing of the window and maintains the position of the window.

Window Casing The trim around the window.

Window Header The top two boards placed horizontally in the frame of a window to support a load above the window.

Window Sill	The lower horizontal member in the frame of a window.
Window Stool	The window sill.
Wrapped Opening	An entryway that is wrapped with drywall and has no trim or door.
Zoning Ordinance	Ordinances determined by a municipality that establish the zoning restrictions and classifications.